The Three Mile Island Crisis

The Pennsylvania State University Studies No. 49

The Three Mile Island Crisis

Psychological, Social,
and Economic Impacts on
the Surrounding Population

by
Peter S. Houts
Paul D. Cleary
Teh-Wei Hu

Foreword by George K. Tokuhata

The Pennsylvania State University Press
University Park and London

Library of Congress Cataloging-in-Publication Data

Houts, Peter S.
    The Three Mile Island crisis.

    (The Pennsylvania State University studies ; no. 49)
    Bibliography: p.
    Includes index.
    1. Three Mile Island Nuclear Power Plant (Pa.)
2. Nuclear power plants—Accidents—Economic aspects—
Pennsylvania. 3. Nuclear power plants—Accidents—
Social aspects—Pennsylvania. 4. Nuclear power plants—
Pennsylvania—Accidents—Psychological aspects.
I. Cleary, Paul. II. Hu, Teh-Wei. III. Title.
IV. Series.
HD9698.U53P45   1988   363.1′79   87-43186
ISBN 0-271-00633-1

Contents

AREA SURVEYED
General Population Studies of Three Mile Island Crisis Impact

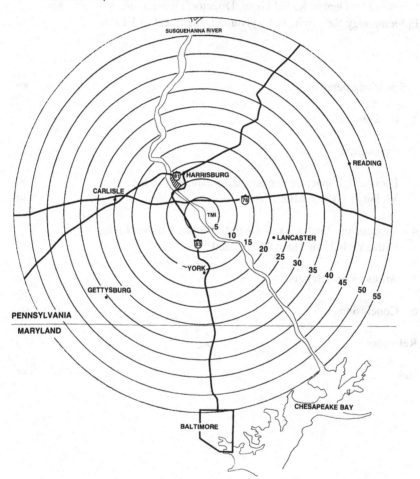

Foreword

The Three Mile Island accident, unlike natural disasters such as floods, hurricanes, earthquakes, or major fires, left no immediate physical alterations in the surrounding environment. On the other hand, it did cause an extensive evacuation of persons living near the facility, substantial short-term economic impact on both individuals and companies in the area, and significant psychological disturbances among persons living in its vicinity. These effects resulted from fears about radiation releases from the damaged reactor and concerns about possible immediate and long-term effects on people's health. In addition, incomplete and often confusing information about radiation releases received by the local residents during and after the crisis further intensified their distress.

Since the March 1979 accident at the Three Mile Island (TMI) nuclear power plant, many studies have assessed its impacts. Compiled and summarized in this book are the results of five related surveys, all aimed at the scientific assessment of the psycho-socio-economic behavior of the residents around the TMI facility. These studies are important because they are based on a randomly selected, large sample of the population (with telephones) around TMI. Because of this, the results can be trusted as being representative of that population. Furthermore, the studies were conducted over a fifteen-month period, making it possible to examine how reactions to the crisis changed over time.

Another strength of this book is that the studies presented here are the products of close interagency cooperation involving collaboration among government (state and federal) agencies, independent research organizations, and universities. Furthermore, the designs and findings of much of the data reported here have been reviewed and supported by the TMI Advisory Panel for Health Research Studies of the Pennsylvania Department of Health, which consists of experts in the different disciplines relevant to the TMI crisis and from various academic and governmental institutions.

The behavioral studies reported in this book represent an important part of the overall health studies, including epidemiologic inquiries, related to the TMI accident. It should be noted that the data presented here are essentially "reported" by individuals residing in the vicinity of the TMI facility. Limitations and qualifications of this type of data, in contrast to epidemiologic data, are duely described by the authors. Since the amount of radiation released during the TMI accident was officially reported to be minimal, the psychosocial impacts of the accident upon local residents may be more important than its effects upon their physical health. As epidemiologic studies of the TMI-area populations are completed, their results can and should be integrated with the results of the behavioral studies to provide a comprehensive interpretation of the impact of the Three Mile Island crisis.

George K. Tokuhata, Dr.P.H., Ph.D.
Director, Division of Epidemiology Research
Pennsylvania Department of Health

Preface

The Three Mile Island Crisis

At approximately 4:00 A.M. on Wednesday, March 28, 1979, an accident occurred at the nuclear generating station at Three Mile Island (TMI), Pennsylvania, involving one of the two reactors at the facility. A series of mechanical and judgmental errors caused the loss of a substantial part of the protective blanket of water in the reactor, and as a result, as much as two-thirds of the nuclear core was uncovered and part of the core melted. In addition, an undetermined amount of radioactive gas was released into the atmosphere (Presidential Commission 1979). It was the most serious accident in the history of nuclear power to that time.

On Wednesday and Thursday following the accident, Metropolitan Edison, the company that operated the Three Mile Island facility, announced that there was no danger to the public, though others interviewed by the media expressed some skepticism. However, on Friday, March 30, the governor of Pennsylvania, Dick Thornburgh, responding to concerns about radiation releases at Three Mile Island, recommended that pregnant women and preschool children leave the area within five miles of the facility. Schools within that radius were also advised to close. Sirens were set off in many communities in the area, and local radio and television stations interrupted programming with announcements of where evacuation centers would be located, who should evacuate, and the likelihood of more people being advised to leave the area. National television soon picked up the story and began almost continuous coverage of the "Three Mile Island Crisis," focusing attention on the possibility of a hydrogen bubble developing inside the damaged reactor, which could lead to an internal explosion and/or to further exposure of the nuclear fuel, which, in turn, could lead to meltdown of the reactor's nuclear core. The news reports emphasized continuing danger not only to the people living near the facility but also to a significant proportion of the population on the northeast coast of the United States. One newscaster intoned, "The danger faced by man for tampering with natural forces, a theme familiar from the myths of Prometheus to the story of Frankenstein, moved closer to fact from fancy."

The crisis quickly became a worldwide news event, and reporters from national and international media converged on Harrisburg, the state capital, only ten miles from the facility, and Middletown, three miles away. For the next five days Three Mile Island was the focus of intense media attention, and both Governor Thornburgh and President Jimmy Carter became actively involved and visited the facility. Though the evacuation advisory applied to only approximately 12,000 persons (families with pregnant women and preschool children living within five miles of the facility), more than 140,000 persons within fifteen miles left their homes for periods ranging from a day to several weeks.

On Monday, April 9, ten days after recommending evacuation, the governor announced that there was no longer serious danger of a meltdown and withdrew his evacuation advisory. Television, radio, and newspaper coverage then decreased, though the media continued, for over a year, to report regularly on the results of various fact-finding commissions as well as charges and countercharges of responsibility for what had occurred.

Eight years later, as this book is being completed, the implications of the TMI accident are still growing. Studies are continuing on possible health effects of the radiation released during the crisis as well as stress associated with living near the facility. General Public Utilities, the company that owned the Three Mile Island facility, has been severely affected economically. For six and a half years following the accident, there were protests and lawsuits concerning restarting the undamaged reactor at Three Mile Island. In addition, more than a thousand persons became involved in legal actions claiming physical and/or mental harm from the accident. Nationally, there have been increased questioning of the safety of nuclear power and increased demands for closer regulation of the nuclear power industry, in part because of the events at Three Mile Island in March and April of 1979.

Features of the TMI Crisis That Affected Its Impact

The TMI accident is especially interesting because it is a prototype of a kind of crisis that is increasingly likely to affect our lives: crises caused by failures of technology. In the not very distant past, the disasters faced by man were primarily "natural," such as floods, hurricanes, earthquakes, and poor crops. These events were so prominent in men's lives that Malinowski (1955) argued that they formed the basis of man's rituals of magic and religion. However, as man attempted to gain control over his environment and to harness the forces of nature, man-made devices increasingly came to be the source of danger. Instead of flood damage being only an "act of God," it became the result of a failed dam, a negligent coal company, or inadequate engineering surveys. Recently, many of the disasters we face have been almost entirely the result of failed technology. The crash of an airplane, the poisonous lead in gasoline, dioxin in the water supply, and nuclear accidents are primarily the result of the failure of machines or of poor judgment or planning on the part of their designers or operators.

The nuclear accident has features that are likely to occur with increasing frequency as technology becomes more complex, more dangerous, and more widespread—that is, where the dangers are difficult to see, measure, or understand; where the public is dependent on experts to assess danger; where blame is directed at those perceived as responsible for the accident; and where the costs of the crisis extend over a long period of time. Each of these features played a significant role in the type and extent of the crisis's impact on persons living in the vicinity of the damaged reactor at Three Mile Island. We discuss these issues in more detail below.

Radiation, which was the principal source of danger from the accident, is especially frightening because it cannot be felt or seen and is not easily understood. It is not possible to tell by one's senses when one is being radiated. Therefore, it was impossible for people living near Three Mile Island to assess the danger directly or to know what protective actions to take. This uncertainty and inability to protect themselves not only raised levels of fear during the crisis but also created feelings of frustration and a sense of helplessness. Furthermore, the effects of the radiation released on people's health, if any, will not be known for many years. This caused continuing uncertainty long after the accident and contributed to the persistence of concerns among persons living near the damaged reactor. In addition, releases of radiation can affect a large geographic area, well beyond the five miles specified in the governor's evacuation advisory. This was a major factor in the extensive evacuations that occurred beyond the five-mile radius.

A second important feature of the TMI accident was that the public was dependent on experts to assess the situation because the average citizen could not easily understand or assess the danger he or she faced. During and after the TMI crisis, the media gave considerable attention to disagreements among experts about the degree of danger present. These disagreements were cited frequently in surveys as playing an important role both in the extensive evacuation during the crisis and in the distress reported by people living near TMI, which persisted for at least a year after the accident.

A third important feature of the TMI crisis was that, because human beings built and operated the power plant, it was possible to blame people for what occurred. This differs from responses to most natural disasters, where the source of the danger is seen as an "act of God" and therefore beyond man's control. Blame for the TMI accident was directed at personnel of Metropolitan Edison Company (which operated the Three Mile Island facility), and this affected both attitudes among persons near the facility and the political effects of the crisis. Following the TMI crisis, attitudes toward Metropolitan Edison were very negative. Many people felt the company was not trustworthy and so opposed their restarting the undamaged reactor; both political and legal efforts were directed toward this end. While these efforts did not directly affect the restart, they did put pressure on the Nuclear Regulatory Commission to be more cautious and rigorous in monitoring not only the Three Mile Island facility but nuclear power plants throughout the nation.

Throughout this book we refer to how these features affected responses to the TMI crisis. In the Conclusions chapter we summarize their implications.

The Scope and Organization of This Book

This book is a record of how people living in the vicinity of Three Mile Island were affected by the accident over the eighteen-month period following the crisis. It summarizes the psychological, social, and economic effects as assessed through 3,649 telephone interviews with persons living up to fifty-five miles from the plant. The first interviews were conducted in July 1979, three months after the accident, and the last were in October 1980, fifteen months later.

Several features of these surveys provide important perspectives on the TMI crisis that are not available in other research publications. One is how the impact of the crisis changed over time. The surveys reported here were conducted at three, nine, and eighteen months after the crisis, and because the later surveys repeated many questions from the earlier surveys, it was possible to examine how responses to the same questions changed over time and how the characteristics of persons who responded in certain ways changed as well. Another unique perspective is provided by the fact that these surveys extended out to fifty-five miles from the facility. They are the only surveys of the population near TMI that randomly sampled households with phones over such a large geographic area. As a result, the findings can provide context for interpreting findings from other studies that sampled only subgroups of the population or where the sampling methods used could have resulted in significant sample bias. Another advantage of sampling from a large geographic area is that it was possible to assess the effects of living close to TMI by comparing responses of persons living near the facility with responses of persons living fifty to fifty-five miles away.

This book was written with three audiences in mind: scholars concerned with the social and political significance of the Three Mile Island crisis, planners concerned with preparing for public emergencies, and social scientists concerned with understanding why people respond as they do to crisis situations. For those interested in the social and political significance of the TMI crisis, this book provides a record of how people living near Three Mile Island felt and acted during and following the crisis. Because of the rigorous sampling methods used, the large geographic areas studied, and the extended time period covered, these surveys are the most complete record available of how the general population living near Three Mile Island responded to the crisis. For the planner concerned with preparing for public emergencies, this book provides detailed data on the numbers and types of people affected by the crisis as well as how their lives were affected. This information can be useful in anticipating the service needs of populations affected by similar crises in the future.

For social scientists, the Three Mile Island crisis is of interest because it is an opportunity to extend knowledge about how people respond to crises and stressful situations in general. We have referred to the extensive literature on evacuations during crises, economic impacts of crises, responses to stress, coping, and attitudes in interpreting the results of the surveys. In each chapter we have included a summary of related work by other TMI investigators as well as comparisons of our and their findings. We have also summarized findings from studies of other crises and discussed how the findings at TMI relate to other work on disasters.

Chapter 1 is intended to provide a context for interpreting the findings reported in later chapters. It includes a recapitulation of events during and following the crisis, with quotations from newspaper articles prior to each of the surveys. These quotes show the context within which each of the surveys was conducted and the public events that people were responding to when they answered the survey questions. Chapter 1 also includes data on the demographic characteristics of people living near TMI at the time of the crisis and how they compare to the population of the United States as a whole. This is intended to help the reader assess the generalizability of the survey results. The final topic in this chapter describes the telephone interviews: when they were conducted, sampling methods, and sample sizes. We also discuss the limitations of telephone surveys to alert the reader to qualifications that should be kept in mind when reviewing study findings.

Chapters 2 through 5 summarize the findings for four different types of impacts: evacuation, economic effects, stress, and attitudes. Each chapter begins with an overview, followed by summaries of related research, including studies by other TMI investigators. The findings of the general population surveys are then presented, followed by a conclusions section in which we discuss general issues raised by these findings and how they relate to those of other investigators.

The final chapter summarizes findings on the scope and persistence of the crisis's impact as well as where differences were found between respondents' reports and objective assessments of impact. This is followed by a discussion of the main features of the TMI crisis that significantly affected its impact. Finally, we discuss the implications of our findings for public planners concerned with preparing for crisis situations and for social scientists concerned with understanding and conceptualizing how people respond to stressful situations.

Acknowledgments

Dr. George Tokuhata, Director of Epidemiology Research for the Pennsylvania Department of Health, has been a strong supporter as well as collaborator in much of the research presented in this book. As director of the Pennsylvania Department of Health programs that assess the impact of the Three Mile Island crisis, he has been the department's spokesperson on many of the controversial issues concerning Three Mile Island. In spite of the many pressures he was under, he always insisted on complete professional independence for investigators both in designing studies and in interpreting findings. Additional help and support have come from members of the Three Mile Island Advisory Panel on Health Research Studies of the Pennsylvania Department of Health, who reviewed both the plans and the results of many of the surveys discussed here.

We appreciate the help of Dr. Cynthia Flynn of Social Impact Research Inc. and Mr. Michael Kaltman of the Nuclear Regulatory Commission (NRC). They used many questions from the Department of Health's July 1979 survey in the NRC TMI survey, which permitted the two studies to complement each other. Dr. Flynn and Mr. Kaltman were also generous in giving us access to the NRC data that was utilized in many analyses reported in this book.

A number of social scientists shared their professional experience and expertise in study design, questionnaire construction, data analysis and interpretation. While they are not responsible for conclusions drawn here, they have contributed substantially to our work. We wish to especially acknowledge the assistance of Dr. David Mechanic, Dr. Morton Kramer, Dr. Elmer Streuning, and Dr. Evan Pattishall. We also want to thank Dr. Bruce Dohrenwend, Dr. Evelyn Bromet, and Dr. Andrew Baum for reading and commenting on a draft of Chapter 4 and Mr. Conrad Six, Mr. Fred Hafer, Mr. Craig Burgraff, Mr. Vincent Butler, and Mr. Harold Piety for providing information on the long-term economic impacts of the TMI accident.

We would like to acknowledge the technical help of the staff of Chilton Research Services, who conducted the telephone surveys that are the focus of this book. Ms. Carol DeGennaro and Ms. Nancy Kreuser contributed to questionnaire design and were also responsible for the efficient, professional manner in which interviews were conducted.

Ms. Nancy Goldberg has provided excellent editorial help in the long process of writing this book. We also received editorial assistance from Mary D. Houts and Dr. Robert Crist. Finally we would like to acknowledge four persons who collaborated with us in earlier publications concerning the Three Mile Island crisis. Their contributions to the findings presented here have been substantial. They are Dr. Kenneth Slaysman, Ms. Marilyn Goldhaber, Dr. Michael Lindell, and Dr. Robert Miller. Dr. Miller's untimely death has saddened all who knew him personally and professionally. We dedicate this book to his memory.

1

Background

Overview

In this introductory chapter we provide background information to help the reader interpret the findings presented in later chapters. First we recapitulate the major public events concerning Three Mile Island beginning with the accident on March 28, 1979, and proceeding through October 1980, the date of the last survey. We include quotes from newspaper articles just prior to each survey to show the context within which people responded to the survey questions. Second, we discuss the demographic characteristics of the population within five miles of TMI and compare them to the nation as a whole. The third section discusses the strengths and limitations of the telephone survey methods used in the studies reported here. And fourth, we describe the timing, sampling, and content of each of the general population surveys discussed in this book.

During the crisis the media emphasized the immediate dangers for persons in the vicinity of the damaged reactor as well as the possibility of radioactivity spreading over large parts of the eastern seaboard. Three months later, prior to the first survey, newspaper articles dealt with who was to blame for the accident, how dangerous the accident had been, and meetings where local citizens called for closing the undamaged reactor at Three Mile Island. In January 1980, prior to the second wave of surveys, there were numerous articles about how the cleanup of the damaged reactor was progressing, but also many that still discussed the dangers of the original accident. This included a just-released report that the TMI accident had come within thirty to sixty minutes of a meltdown. By October 1980, however, the newspaper articles on TMI focused almost entirely on the cleanup program, including both its problems and accomplishments. This change in focus from the danger of the accident to the cleanup efforts is paralleled by gradually decreasing levels of concern and distress reported in the chapters on attitudes and stress.

The comparison of demographic characteristics of people living within five miles of TMI with those of the nation as a whole showed that the TMI population is similar to the nation in age of heads of households, family size, and church attendance. However, there are fewer blacks, more married, more who completed high school, more families with incomes greater than $10,000, more home owners, and more blue-collar workers than for the nation as a whole. Overall these findings suggest that people living in the immediate vicinity of TMI may be somewhat more conservative than is the case nationally.

The strengths and limitations of telephone surveys should be kept in mind when interpreting the findings presented in later chapters. Random digit dialing, the method used to contact respondents, is intended to provide a representative sample of households with telephones. However, the final sample can be biased because of higher refusal rates among certain groups of the population, because of undersampling of people who work during the evenings and weekends, and because of the exclusion of households without telephones. Other factors that may affect the validity of the results include respondents' conscious and unconscious biases; their inability to correctly remember events, feelings, or attitudes; and the tendency that many people have to try to appear consistent or to please interviewers.

Another issue that should be kept in mind when interpreting study findings is the distinction between statistical and practical significance. The former refers to whether an association or difference between groups is likely or unlikely to have occurred by chance. The latter refers to whether the association or difference is large enough to be of practical significance. With large samples, as were used in the TMI general population surveys, small differences between groups may be statistically significant, but still not be large enough to be of practical significance. Both statistical and practical significance should be considered when interpreting results.

Recapitulation of Events Concerning TMI over the Eighteen-Month Period Discussed in This Book

The principle source for the material presented in this section is the *Three Mile Island Sourcebook* (Starr and Pearman, 1983), which summarizes both local and national news reports about Three Mile Island beginning in November 1966, when announcements were first made about constructing the nuclear power plant, and continuing through March 28, 1983, the fourth anniversary of the accident. Our presentation focuses on the public rather than the technical events at the facility because we are concerned with how people responded to publicly available information provided by the parties involved and the news media.

The Crisis Period

The accident at reactor #2 at Three Mile Island began at approximately 4:00 A.M. on Wednesday, March 28, 1979. The governor of Pennsylvania, Dick

Thornburgh, was notified about what had happened at 7:45 A.M., and President Jimmy Carter was informed at 9:00 A.M. Also at 9:00 A.M., the Associated Press released a national bulletin stating that a general emergency had been declared at the Three Mile Island nuclear power plant but that no radiation had been released. Later in the day both John Herbein, the vice president of Metropolitan Edison (Met Ed), the utility that operated the plant, and the lieutenant governor, William W. Scranton III, held press conferences stating that there was no danger to the public's health (though Scranton was also reported as saying that Met Ed had given misleading information about what had occurred). On the next day, Thursday, Governor Thornburgh held a press conference where he reaffirmed that there was no hazard to the health of the general public at TMI.

Friday, March 30, is the day that most local residents associate with the crisis. At 8:00 A.M. a higher than expected radiation reading was recorded above a vent stack at Three Mile Island. This led some staff of the Nuclear Regulatory Commission in Washington, D.C., to recommend evacuation of people living near the plant, but an evacuation order was not issued. However, later in the morning Governor Thornburgh, responding to concerns expressed by his secretary of health and by the chairman of the Nuclear Regulatory Commission about the possible health effects of further radiation releases, recommended that people within ten miles of TMI stay indoors. Later, at 12:30 P.M., he recommended that pregnant women and preschool children living within five miles of TMI leave the area. He also recommended that schools within five miles of the plant be closed.

After the governor's advisory, a number of local events alerted people to the possible danger and, at the same time, magnified the perception of danger. Sirens were set off in many townships in the area to attract attention to the problem. Local radio and television began almost continuous coverage, which included repeatedly playing the governor's statement, as well as interviews with civil defense and other officials about what might happen. Some stations even broadcasted lists of what people should take with them if they were to leave in a rush, as well as the locations of the evacuation centers that had been designated for pregnant women or women with young children. There was a sense of urgency and seriousness to these broadcasts, communicating to many a level of danger exceeding the formal statements issued by the governor and other officials. That afternoon, many parents in the vicinity of TMI took their children out of school. For the children that remained, there was often a sense of confusion and uncertainty about what was happening. An anecdote shows how some children felt at that time. A twelve-year-old boy scratched the following note for posterity on his bedboard just before leaving on March 30: "David Houts and family left this house on March 30, 1979, because the nuclear plant was leaking radioactive materials. He left to Connecticut where his grandmother lived." The boy who wrote this note, like many people who evacuated, was not sure he would ever return to his home. He wanted to leave a record for the unknown people in the future who would visit his house in what he and many other evacuees thought might become a radiation wasteland.

On Friday evening and continuing through the weekend, national television and radio provided continuous coverage of the situation at Three Mile Island. This included discussions of how nuclear power plants work, the meaning of hydrogen bubbles in the reactor vessel, and projections of where fallout would occur from the radioactive clouds that could result from a meltdown at Three Mile Island. Officials, scientists, and politicians were interviewed frequently, and the television and radio announcers continued to communicate a sense of urgency and danger by the manner in which they presented material. Nimmo (1984) has analyzed the content of the presentations of the TMI crisis by the three major television networks and found that they emphasized the fearful and dramatic aspects of what occurred.

On Friday evening, Harold Denton, a member of NRC's emergency management team and President Carter's personal representative at the site, held a press conference with Governor Thornburgh. Mr. Denton stated that a core meltdown was possible, but only remotely so. Thereafter, Mr. Denton became the primary spokesperson for the NRC. Other groups continued to make their own (sometimes contradictory) news releases.

On Saturday and Sunday there was considerable discussion at press conferences by Mr. Denton and others about the possibility of a hydrogen bubble forming in the reactor vessel. If such a bubble were to grow to sufficient size, it would displace the cooling water, expose the core, and possibly lead to a meltdown. Mr. Denton continued to state that, while such a sequence was possible, it was unlikely. Governor Thornburgh repeated that there was no need for a full evacuation at that time. On Sunday, April 1, President Carter visited Three Mile Island and concurred with Mr. Denton's and Governor Thornburgh's views of the situation.

Beginning on Monday, April 2, news releases by the NRC and the governor's office projected a more positive and hopeful tone. It was stated on Monday that the size of the hydrogen bubble was decreasing and, on Tuesday, that the situation at the plant was stable and there was no longer the threat of a hydrogen explosion. From April 4 through 6, Mr. Denton emphasized that only minor problems were occurring, though the governor's evacuation advisory for pregnant women and preschool children remained in effect and schools within five miles of TMI were advised to stay closed. Many of the schools beyond the five-mile radius, which were closed early in the week, began to reopen on Thursday, April 5. On Monday, April 9, Governor Thornburgh lifted the evacuation and school advisory, ten days after it was issued. On April 17, Mr. Denton left the site to return to Washington. It was announced on April 27, 1979, that the plant was placed on natural circulation cooling.

As can be seen by this summary of events, the "crisis" period, when the

4

public was told that there was serious danger, peaked from Friday through Sunday, with a gradual decrease in levels of danger reported over the following month.

Events Prior to the Telephone Surveys

Public attention to the situation at Three Mile Island did not end with Governor Thornburgh's lifting of the evacuation advisory. A review of the *Guide to Periodic Literature* for the period from March through December of 1979 showed more than a hundred citations of articles concerning Three Mile Island. National television newscasts regularly included reports concerning TMI. In June and July of 1979, when the first surveys were conducted, there were TMI-related articles almost daily in the local and national press. Sample citations from the *Three Mile Island Sourcebook* are listed below. They cover the period from June 1, 1979, through July 15, 1979, immediately preceding and including the times when the first surveys were conducted.

(6/1/79) "NRC Scrapped Evacuation." Denton had recommended an evacuation on 3/30, but rescinded it within an hour. . . . He also stated that the NRC staff was somewhat complacent about reactor safety before the accident.

(6/9/79) Mayor Reid [of Middletown, within five miles of TMI] blames the NRC for the TMI affair. He contends the plant was licensed before it was ready and that there were no inspectors at a plant with the potential to kill thousands.

(6/13/79) James H. Fisher, Executive Director of Emergency Health Services of South Central Pa., contends "mass confusion" among emergency officials occurred in the wake of TMI.

The director of the Bureau of Radiological Protection [a Pennsylvania state agency] . . . stated that there was never a danger or possibility of hydrogen explosion, and that the evacuation of pregnant women and preschool children was unnecessary.

(6/21/79) In a meeting attended by more than 250 Middletown residents, the Borough Council was criticized for not passing a resolution calling for the closing of Unit I [the undamaged reactor at Three Mile Island].

Forty thousand residents within a five mile radius of TMI are being surveyed to find the effects of low level radiation on health.

(6/22/79)   The NRC says we will never know how much radiation escaped from the TMI plant because the levels exceeded the abilities of the plants to measure them.

(7/3/79)   Expression of the tension and emotion felt at a town meeting in Middletown is described. Congressman Ertel and the NRC's Victor Gilinsky fielded questions from the public. . . . The reporter determined that Gilinsky was evasive in answering.

(7/6/79)   Investigators are seeking to discover why the government was not immediately informed that the core of the TMI nuclear reactor began disintegrating within a few hours after the start of the accident. They ask whether there was a cover-up.

(7/12/79)   Editorial: "We Will Not Be Lulled." The editor alleges that people living under the ominous shadow of TMI do not want public utilities deciding how much radiation is safe for us, nor how much tritium or other radioactive substances we must accept in our drinking water.

Six months later, at the time of the second wave of surveys, Three Mile Island continued to be an important news item both locally and in the national media. One of the major events since the last survey was the issuing of the report of the Presidential Commission on Three Mile Island, which criticized many aspects of how the crisis had been handled. A sample of citations from the *Three Mile Island Sourcebook* for the month of January indicates what people living near Three Mile Island were reading in their local newspapers:

(1/2/80)   The NRC states that Met Ed's program for cleanup of TMI has deficiencies and needs upgrading to assure the safety of its workers.

(1/5/80)   "City Settles Suit Over TMI Water" . . . the NRC also agrees to ban the dumping of radioactive water into the river until 1982, or until an EIS [environmental impact survey] is completed.

(1/10/80)   "Nuclear Peril Minimized" . . . Dr. R. Linneman said that the "likelihood of danger from a Nuclear accident is minimal."

(1/15/80)   The NRC set forth new regulations to prevent release of radiation during a meltdown. This reverses an earlier position that a meltdown was too remote. Following TMI, the NRC staff concluded that chances of a meltdown are higher than previously expected.

6

... claims of $1.3 million ... have been paid to evacuees who left homes at the time of the TMI crisis. This is the largest single payout in the history of Nuclear insurance.

(1/23/80)   Six volunteers are training to become the first humans to enter the containment building at TMI.

The NRC imposes a $155,000 fine on TMI facility.

(1/25/80)   "TMI Accident Only 60 minutes from a Meltdown." The primary conclusion of the Rogovin Panel is that the TMI accident came within 30 to 60 minutes of a meltdown.

(1/29/80)   Pa. DER [Department of Environmental Resources] urges the cleanup of TMI as the radioactive elements within Unit II are viewed as a long term threat.

In October 1980, when the last of the telephone surveys was conducted, articles in the local press concerning TMI had switched their focus from the accident to the cleanup. Since January of 1980, when the last survey was carried out, the anniversary of the accident had occurred, resulting in a good deal of national media coverage, and krypton gas had been vented from the containment building housing the damaged reactor. Controversy had surrounded this venting. While most experts agreed that it posed no health hazard, this view was not universally accepted. Nonetheless, the venting was carried out without a great deal of public protest. Selected citations from the *Three Mile Island Sourcebook* for October 1980 follow:

(10/1/80)   A TMI official charges that conflicting Pa. and federal orders are hampering the cleanup of TMI.

TMI officials indicate that the problems in the Unit I control room are being corrected.

(10/3/80)   GPU [the parent corporation for Metropolitan Edison Company, which runs TMI] is now a candidate to be the first major utility to go bankrupt since the depression. . . .

A report in a Science magazine article states that TMI decontamination might be stalled indefinitely if no permanent disposal site is found for the radioactive wastes.

(10/12/80)   Met Ed takes reporters on a tour of the undamaged Unit I reactor at TMI.

(10/17/80)   A five man team enters Unit II building and performs the first maintenance since the accident.

(10/20/80)   A decision concerning the reopening of Unit I is expected in several months. In preparation, TMIA [TMI Alert, a local organization] has already written a 32 page training manual in methods of nonviolent protest and civil disobedience.

(10/23/80)   "In Middletown, Election Issue is Economy Not TMI." The economy is the key issue as derived from interviews with area residents.

(10/29/80)   A NRC investigation finds no evidence to support charges made by TMIA [TMI Alert] concerning lax maintenance and noncompliance with regulatory requirements at TMI.

(10/31/80)   The results of a NIMH [National Institute of Mental Health] funded study indicated that 25% of the mothers [living near TMI] . . . showed clinical levels of depression or anxiety during the year following the accident.

Demographic Characteristics of the Population within Five Miles of TMI Compared to National Norms

One question that naturally occurs when reviewing data on how people living near Three Mile Island responded to the crisis is: How generalizable are these results to other populations? One factor affecting the generalizability of results is the degree of similarity between this group and comparison groups; the greater the similarity, the more likely it is that results are generalizable.

Table 1.1 shows the demographic characteristics of the 692 persons interviewed in the Pennsylvania Department of Health (PDH) July 1979 survey, conducted three months after the TMI accident. Random digit dialing was used, so this population is a random sample of households with phones within the area. The national comparison data were obtained from the 1980 census. The PDH sample is similar to the national population in terms of average age of heads of households, average family size, proportion of the population with a Hispanic background, and the proportion who attend church weekly. However, it differs with respect to the proportion of the population that is black (1.0% vs. a national figure of 10.2%) and in the percent married (74% vs. 59% nationally). The differences in percent married are due, in part, to different sampling methods, not necessarily to actual population differences. The census

8

Table 1.1
*Demographic Characteristics of Pennsylvania Department of Health
Sample Compared to National Population*

	PDH survey	National characteristics*
Mean age, head of household	42.4	45.7**
% Married	73.6	59.2
Mean family size	3.1	2.9
% Owning homes	77.8	64.0
% Completing high school	74.3	68.0
% With income over $10,000	80.0	72.6
Occupation		
% White collar	22.5	49.6
% Blue collar	69.0	33.9
% Farmer	1.4	3.0
% Other employed	7.1	13.5
Ethnicity		
% Black	1.0	10.2
% Hispanic	4.0	4.7
Religion		
% Catholic	16.8	27.0
% Protestant	73.3	61.0
% Jewish	.4	2.0
% Other	5.2	4.0
% None	3.6	6.0
% No answer	.7	
% Attending church at least weekly	40.0	40.0

*"National characteristics" calculated from 1980 census data. Since question format and categories were rarely precisely equivalent, all comparative data should be regarded as approximate.
**Estimated from group data.

data include all persons over fourteen years of age in assigning marital status categories, while the PDH survey includes only male and female heads of households eighteen years of age or older in the sample. Therefore, single persons living with parents would be part of the census counts but would not be present in the PDH sample.

Differences between the PDH sample and national norms are also evident when one looks at education, occupation, income, and religious affiliation. The study population is somewhat better educated (74% completed high school, compared to 68% for the U.S. population), is more likely to have a total family income greater than $10,000 (80% vs. 73%), but markedly more likely to be employed in a blue-collar occupation (69% vs. 34%). This predominance of blue-collar workers is partly due to the presence of a steel mill and other manufacturing facilities in the vicinity. However, it should also be noted that the criteria for classification into blue or white collar differed slightly between this study and the census. The proportion of respondents in each category of religious affiliation was generally similar to national data, except that there were fewer Jews and people listing their affiliation as "none." Also, the proportion of the population who own their homes is almost 14% higher than the national average.

One feature of the PDH sample not included in Table 1.1 is that 42% stated that their ethnic background was "Pennsylvania Dutch," meaning that they were descended from German and Swiss immigrants who came to the area in the late seventeenth and early eighteenth centuries. The term "Dutch" is an anglicizing of "Deutsch" (for German). It should be noted that Amish or Mennonites, who are sometimes identified with the term "Pennsylvania Dutch," are, at most, a small part of this group. While there is no data on whether respondents were Amish or Mennonites (though no Amish would be represented since they do not have phones), these groups constitute a small percentage of the persons who consider themselves "Pennsylvania Dutch."

In summary, the population who responded to the PDH survey is very similar to the national population in age of heads of household, marital status, and church attendance. There are differences in ethnic composition, education levels, occupations, and home ownership. These differences are generally in the direction of the TMI population's having slightly higher socioeconomic status, which, in combination with the "Pennsylvania Dutch" ethnic origin of a large percent of the population, suggests that the population within five miles of TMI is likely to be somewhat more conservative than the nation as a whole.

The Five Telephone Surveys

The five surveys discussed here share several characteristics: 1) they were all conducted by telephone, 2) they all utilized random digit dialing to locate respondents, and 3) they were all carried out by the same contractor, Chilton Research of Radnor, Pennsylvania. Comparability across interviews was maximized by utilizing the same methodology and the same organizations to conduct the interviews.

The advantages of telephone interviews include relatively low cost per interview and the ability to monitor the interviews as they take place to insure that the interviewers are explaining the questions clearly and are not leading the respondents to give certain responses. Such monitoring was a part of all surveys reported here. Furthermore, random digit dialing insures that a representative sample of households with telephones (estimated at over 90% of households) will be taken. However, there are also important limitations to this methodology that should be kept in mind when interpreting findings. First, households without telephones are not included, and people living in such

10

households are likely to be different from the population as a whole. Specifically, they are likely to have very low incomes, and as a result, this segment of the population will be underrepresented in the sample. A second problem concerns the inability to interview people from all the households that were called. One reason for failing to obtain an interview is that people were not home when the interviewers called. Since most attempts to call were on evenings and weekends, it is likely that people who work during those times are underrepresented in the sample. In some cases, people were reached but refused to be interviewed. People who refuse to be interviewed may be different from those who agree, and to the extent to which this is the case, the final sample is not representative of the entire population.

A final methodological issue to be considered in interpreting the findings is the distinction between "statistical" and "practical" significance of results. If a finding is reported as statistically significant, it means that it is unlikely (less than five times in one hundred) to have occurred by chance. Practical significance, on the other hand, refers to whether a difference or association is large enough to make a practical difference. An effect may be statistically significant, especially when the sample sizes are large, as they are in the surveys reported here, but not big enough to be of practical use. For example, in Chapter 2 we report that the evacuation rate at different distances from TMI ranged from 66% to 1%. These differences are statistically significant and, furthermore, are large enough to have important practical implications as well. On the other hand, education level of the head of household was also associated with evacuation, but this trend is much less pronounced than for distance and therefore is less likely to be of practical significance. Of course many factors can determine practical significance, but strength of the association is often one of the most important. Another, related issue is the degree to which all the independent variables together are associated with a dependent variable. For example, the fifteen variables (including distance) used to predict evacuation in the NRC survey accounted, together, for only 27% of the variability in whether or not people evacuated. This means that 73% of this variability is due to factors other than those that were studied. Therefore, public officials concerned with planning for crises similar to TMI should be aware that other factors, in addition to those studied here, need to be considered when planning how to direct evacuations.

Throughout this book we refer to the various surveys in terms of their sponsors as well as by the date of the survey. PDH refers to the Pennsylvania Department of Health, NRC to the Nuclear Regulatory Commission, and RWJ to the Robert Wood Johnson Foundation. The five surveys are summarized in Table 1.2 in the order in which they were conducted. A more detailed discussion of the topics covered and methodology used in each of the surveys is reported in publicly available publications referenced in the bibliography.

11

Table 1.2
Summary of the Five Telephone Surveys

Survey date	Survey sponsor	Initials used in text	No. of respondents	Area from TMI sampled (miles)	Comments
July 1979	PA Dept. of Health (PDH)	PDH July 1979	692	0–5	Random sample of households with phones within a 5-mile radius of TMI. Response rate was 75%.*
July 1979	Nuclear Regulatory Commission (NRC)	NRC July 1979	1,504	0–55	0–15 mile samples were from phones anywhere within a 15-mile radius of TMI. 16–55 mile samples were from phones on the north, south, east, and west transects. Response rate was 76%.*
Jan. 1980	PA Dept. of Health (PDH)	PDH Jan 1980 (panel)	403	0–5	All respondents had been interviewed in the PDH July 1979 survey. Response rate was 96%.*
Jan. 1980	Robert Wood Johnson Foundation (RWJ)	RWJ Jan 1980	550	0–55	Sampling procedures same as in NRC July 1979 survey. Response rate was 82%.*
Oct. 1980	PA Dept. of Health (PDH)	PDH Oct 1980	500	0–5 and 41–55	0–5 mile respondents were sampled from Dept. of Health TMI registry.** 41–55 mile sampling procedure was the same as the NRC July 1979 survey. Response rate was 75%.*

*Response rate is the percent of selected households with phones that were interviewed.
**PDH started a registry of local residents within five miles of TMI shortly after the accident. It included an estimated 98% of persons living in the area.

2

Evacuation

"I've gone through fire, and I've gone through flood, [but] this radiation, you can't see . . . and I guess that's why we [left]." (Houts et al., 1980, chpt. 4, p. 1)

Overview

Evacuation behavior during the Three Mile Island accident was studied using data from two primary sources. One was the Nuclear Regulatory Commission survey conducted in July 1979 among persons living within fifty-five miles of Three Mile Island. The other source of data was the survey conducted by the Pennsylvania Department of Health in the same month among persons living within five miles of the TMI reactors.

The data from these surveys indicate that, although only about 1% of persons living more than forty miles from the reactor evacuated, 66% of the households within five miles of TMI had at least one person evacuate. Most of the evacuations occurred on Friday, three days after the accident, and about half stayed away for at least five days. Most persons who evacuated stayed with family or friends, but the distances traveled were substantial. Half of the evacuees reported traveling more than ninety miles.

Perceived danger was the reason most respondents gave for evacuating, but about 80% of respondents cited confusing information as a reason for leaving and almost as many reported that they left the area to avoid forced evacuation. More than half of the respondents said that they left to protect children from danger. When persons who did not evacuate were asked why they stayed in the area, the most common reasons given were that they were waiting for an evacuation order, that they felt the situation was in God's hands, or that they saw no danger. Surprisingly, only about 10% of the persons who stayed in the area said that they did so because they had nowhere to go, and less than 5% said that they stayed because of lack of transportation.

The persons most likely to leave were persons who had characteristics mentioned in the governor's evacuation advisory. Living close to the reactor and having young children in the household were the strongest predictors of evacuation. In the survey of persons living within five miles of the reactor, persons in larger households were more likely to evacuate, and the data from both surveys indicate that respondents with more education were somewhat more likely to evacuate. However, it should be noted that the regression models developed to explain evacuation behavior explained relatively little of the variance, indicating that factors other than those studied played important roles in determining whether or not a person or household evacuated.

Two features of the TMI evacuation are in contrast to what has been found about evacuations in natural disasters. First, disagreement among experts was frequently cited as a reason for evacuation during the TMI crisis, while in many natural disasters such disagreement is often cited as a reason for not leaving. The reason for this difference is probably that, in a crisis where the danger cannot be seen or heard and where the public is dependent on experts to assess danger, disagreement among experts is disturbing and therefore likely to lead to taking the conservative course of evacuating. On the other hand, in natural disasters people often feel that they can judge the danger themselves and often overestimate their ability to deal with it. Therefore, when experts disagree, people often rely on their own judgment, which leads to underestimating the real degree of danger. The second difference between evacuations at TMI and in natural disasters is that many more people left than were advised to do so during the TMI crisis, while the opposite is often true in natural disasters. We attribute this to the characteristics of radiation as a source of danger. Its association with cancer and atomic bombs (which makes it especially fearful), the fact that it can spread over a large geographic area and that it cannot be directly sensed, and the public's dependence on experts who disagreed with each other led many people who were not advised to leave to do so nonetheless.

Background

The practice of evacuation is as old as human history. Twenty-five hundred years ago, for example, Egyptians had to evacuate annually to escape the flooding Nile. Hans and Sell (1974) compiled a list of more than five hundred natural and man-made disasters that required evacuation during the 1960s and early 1970s. Their study showed that an average of almost ninety thousand persons per year evacuated from their homes because of natural and man-made disasters and that about 25% of these evacuations were due to man-made disasters. The number of evacuees ranged from a minimum of 25 to a maximum of 501,000 persons, with a median of less than 1,000 persons.

The consequences of most types of natural disasters are familiar to people because of their frequent occurrence and the existence of historical records. Technological disasters, on the other hand, involve less familiar consequences because they occur relatively infrequently and involve forces that many people do not understand. The 1979 nuclear power plant accident at Three Mile Island is a case in point. Nuclear accidents have occurred so infrequently and estimating the consequences of exposure to radiation is so complex that the hazards associated with different levels of radiation exposure are not clearly established, especially for low levels of exposure. As a result, before the TMI accident people living near nuclear power plants were not well informed about or prepared for the potential serious health hazards. This lack of knowledge concerning possible catastrophic consequences and uncertainty about the seriousness of the outcome, which are more characteristic of technological than of natural disasters, are two of the reasons that people's responses to the TMI crisis may be different from what would be expected in most natural disaster situations.

Review of Recent Evacuations

Stallings (1984), in a review of the evacuation literature, made the following generalizations based on earlier research on this subject. Most people evacuate as families, usually finding accommodations with friends and relatives. The average distance traveled by evacuees is thirteen miles, with a range from one-quarter mile to eighty miles. People of higher socioeconomic status are more likely to evacuate than persons of low socioeconomic status. People decide to leave when they believe that the threat is real and serious. And, finally, people are more likely to feel that the threat is real when emergency announcements are consistent and frequent.

Perry, Lindell, and Greene (1981) examined the evacuation experiences of four flooded United States' communities: Fillmore in the Southwest, Valley in the Midwest, and Snoqualmie and Sumner in the Northwest. All four sites experienced riverine flooding between December 1977 and March 1978. Anywhere from one hundred and fifty to seven hundred households were affected by the flood at each site. Approximately two hundred households were sampled in each community to study factors that affected household evacuation decisions. It was found that an individual's perception of risk was an important factor in how he responded to an evacuation warning. If he perceived the threat as real, he was likely to evacuate in response to the warning. The survey data also indicated that people warned by an authority are more likely to perceive the threat as real than people warned by other sources. Finally, prior knowledge of some standing evacuation plan, or being informed of such a plan by a third party, encouraged families to evacuate.

Perry, Lindell, and Greene (1981) also reported that prior warning and visible high water were the reasons most frequently cited for evacuating. The majority of evacuees used a family vehicle for transportation, and they tended to take shelter in relatives' homes rather than in public evacuation facilities. In each of the surveyed communities, almost half of the evacuees expressed concern about looting, but relatively few people cited this fear as a reason for not evacuating. Of the sociodemographic factors examined, age correlated with evacuation in all four study sites. As age increased, the probability of evacuation also increased, which is the opposite of findings reviewed by Stallings (1984). The relationship between age and evacuation is, therefore, uncertain.

Studies have also been carried out on two major natural disasters since 1970: Hurricane Agnes in 1972, with its resultant flood damage at Wilkes-Barre and Kingston, Pennsylvania; and the 1980 volcanic eruption of Mount Saint Helens in Washington. Richard (1974) reports that in the Agnes flood more than 100,000 people were directly affected, 80,000 of them evacuated, and 20,000 to 30,000 permanently dislocated. Melick (1976) found that the mean length of evacuation was 31 days and the median was 31.5 days.

Mount Saint Helens erupted on May 19, 1980. The eruption destroyed 150 square miles of forest, killing both vegetation and wildlife. Sixty-eight people were listed as killed or missing. The area's population is relatively small, approximately fifteen hundred, with most residents involved in some aspect of the logging industry. Perry's (1981) study of evacuation during that crisis found that sirens and public address systems were used by county officials to alert residents of the initial eruption. These techniques were quite successful since almost all residents evacuated to predesignated areas before the eruption occurred.

Other Studies of Evacuation at TMI

The accident at Three Mile Island has been studied by several investigators. Flynn (1979) utilized the NRC's July 1979 data to provide a descriptive account of evacuation behavior extending to fifty-five miles from the reactor. Many of the findings presented in this book are from the same NRC data set. The NRC sample used random digit dialing to obtain a representative sample of homes with phones and obtained a 76% response rate. Zeigler, Brunn, and Johnson (1981) relied on a mailed survey conducted by the Michigan State University group, as well as the NRC data, to describe the geographic aspects of the evacuation. The Michigan State survey included 150 respondents within twenty-five miles of TMI. Their response rate was 56%. Barnes, Brosius, Cutler, and Mitchell (1979) also conducted a mailed survey of persons listed in phone books and living in the vicinity of Three Mile Island. They had a response rate of 39%, with respondents who were predominately male home-owners with more than a high-school education. Kraybill (1979) conducted a telephone interview with 375 persons (selected from telephone directories) living within fifteen miles of TMI approximately a week following the accident. The response rate was 82%. Smith (1979) interviewed 123 persons selected from the Middletown telephone directory. (Middletown is within five miles of Three Mile Island.) Interviews were conducted over a three-week period beginning three days after the accident. The response rate was 92%.

All of these surveys involved some sample bias. Those that used telephone directories missed households without listed phones or with no phones. Random digit dialing missed households without phones, and those surveys with low response rates have unknown sources of bias. The Kraybill study, which was conducted while many evacuees were still away, is likely to underestimate evacuees, especially those who stayed away the longest. It is difficult to compare the estimates of evacuation during the crisis since some investigators reported households, some reported respondents, and some reported respondents in combination with other members of the household. In addition, not all investigators studied the same geographic areas.

Several investigators reported on evacuation within five miles of TMI. Flynn estimated that 60% of all persons within five miles of TMI evacuated and that 66% of households contained at least one evacuee. Zeigler et al. reported that 55% of their respondents had evacuated; Kraybill reported that 51% of his sample left during the crisis; and Smith reported that 57% of his respondents had evacuated during the crisis. The Pennsylvania Department of Health conducted a survey of approximately 98% of persons living within five miles of TMI several months after the crisis and reported that 58% of all persons over seventeen years of age living in the area had evacuated during the crisis. While the groups studied by the different investigators are not completely comparable, there is a consensus that evacuation was within the 50 to 60% range.

Houts et al. (1984) have presented a theoretical analysis of evacuation behavior at Three Mile Island. They utilized the "protective action decision model," which uses four types of variables to predict when people take protective actions: perceived severity of the hazard, perceived susceptibility to the hazard, barriers to taking protective actions, and costs of taking protective actions. This model has been applied to a wide range of protective behaviors and includes Rosenstock's (1966) "Health Belief Model," which has been extensively used to study health-related behaviors, and Perry et al.'s (1981) model to explain responses to threat from disasters. The NRC survey following the Three Mile Island crisis included questions related to three of the protective action variables: 1) perceived severity of the hazard (how serious a threat the respondent felt the TMI plant was to him/herself and family's safety at the time of the accident), 2) susceptibility to the hazard as indicated by the similarity of the respondent to those persons whom the governor advised to evacuate (distance from TMI, whether someone in the household was pregnant at the time of the crisis, and whether there was someone under six years of age in the household), and 3) barriers to evacuation (e.g., could not leave job, lack of transportation, too sick to travel). Discriminant analysis using the severity and susceptibility variables showed that all were significantly related to evacuation and that the model correctly classified 73% of respondents in terms of whether someone in the household evacuated during the crisis. Additional analyses utilizing the barrier variables indicated that they too played a role in evacuation behaviors during the TMI crisis.

Another model for explaining evacuation during the TMI crisis has been proposed by Johnson and Zeigler (1986). Their model utilized locational, life-cycle, social-status, attitudinal, and risk-perception variables to predict TMI evacuation. Using the same NRC survey as used by Houts et al., they predicted whether someone in the household evacuated from: 1) location in relation to the plant (including both distance and direction), 2) stage in the family life cycle (including age of the household head, marital status, young children in the home, and anyone pregnant in the home), 3) social class (years of education), and 4) the respondents' attitudes before and during the crisis (the advantages vs. the disadvantages of nuclear power in general as well as the TMI plant in particular, and how serious a threat the respondent felt the TMI plant was to him/herself and family's safety at the time of the accident). Their model correctly classified 71% of respondents in terms of whether at least one person in a household evacuated during the crisis.

Since the Houts et al. model and the Johnson and Zeigler model used many of the same independent variables to predict the same dependent variable, it is not surprising that their results are similar. Another way of looking at both sets of findings is that, since only 37% of all NRC respondents said that someone in their household evacuated, one could correctly classify 63% of the respondents by simply predicting that no one evacuated. Therefore the Houts et al. model and the Johnson and Zeigler model improved on a prediction based on the average evacuation by 10% and 8% respectively. There is an important limitation that should be noted for both studies, namely, that respondents' reports of perceived threat from TMI and other attitudes were assessed after the evacuation had taken place. Therefore, it is possible that respondents' reports of these attitudes were adjusted retrospectively to be consistent with their evacuation behaviors, as would be predicted by Festinger's theory of cognitive dissonance (1957). If so, then both studies may have overstated the predictive power of the models.

This chapter utilizes data from several sources to provide a comprehensive description of the evacuation decision during the Three Mile Island crisis. The two primary sources are the Nuclear Regulatory Commission survey in July 1979, which polled people living within a fifty-five-mile radius of Three Mile Island, and the Pennsylvania Department of Health survey of the same month, which dealt only with those persons living within a five-mile radius.

Scope of the TMI Evacuation

It is important to keep in mind that there are several elements that facilitated evacuation behavior, including the facts that the weekend was about to begin

FIGURE 2.1

PERCENT OF HOUSEHOLDS WHERE AT LEAST ONE PERSON EVACUATED DURING THE TMI CRISIS AT DIFFERENT DISTANCES FROM THREE MILE ISLAND*

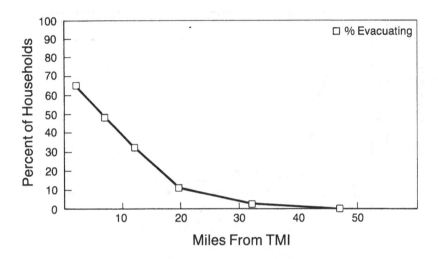

*The six plotted points represent the percent of adults evacuating within the following areas as measured from Three Mile Island: 0-5 miles, 6-10 miles, 11-15 miles, 16-25 miles, 26-40 miles and 40-55 miles.

and the weather was mild. Nonetheless, the extent of evacuation is impressive. According to the NRC data, the total number of persons evacuating within a fifteen-mile radius of the plant is estimated at about 50,000 households or 144,000 individuals (Hu and Slaysman, 1984; Flynn, 1979).

The extent of household evacuation at different distances from TMI is shown in Figure 2.1. This plot shows that, within five miles of TMI, 66% of households had at least one person evacuate, but that forty to fifty-five miles away only 1% of households had someone evacuate. There are at least two possible reasons why distance is so strongly related to evacuation. First, the governor, in his advisory concerning evacuation, had specified a five-mile limit, thereby implying that distance was an important element in determining safety. Second, people probably realized that radiation released into the atmosphere would be likely to dissipate with distance. This very pronounced relationship between distance and response to the crisis is repeated with many other measures of impact discussed in this book.

FIGURE 2.2

LEAVING AND RETURN DATES FOR PERSONS WHO EVACUATED BECAUSE OF THE THREE MILE ISLAND CRISIS

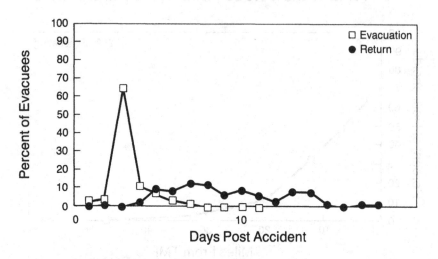

Departure and Return Dates

When people left and returned is an indicator of how concerned evacuees were about the situation at Three Mile Island. Leaving earlier and staying away longer are both indications that evacuees took the warnings of possible danger very seriously. Figure 2.2 shows departure and return dates for persons living within five miles of Three Mile Island. These findings are based on the PDH July 1979 survey.

Two matters of interest are evident from Figure 2.2. The first is that only a small percent of evacuees (8%) left on either Wednesday, March 28, the day of the accident, or Thursday, March 29. The third day after the accident, March 30, is when most (66%) evacuations took place, with sharp drops subsequently to 12, 8, 4, 2, and 1%. These findings clearly indicate that events on Friday were the major stimulus to evacuation and that most people responded quickly. As mentioned earlier, the fact that the weekend was beginning also made it easier to respond to any perceived danger by evacuating.

A second interesting result shown in Figure 2.2 is the remarkably even rate at which people returned. From the fifth to the fourteenth day after the accident, the return rate mostly ranged from 7 to 13%. The mean length of time away from home was between four and five days. Approximately half the evacuees were away for more than five days. This pattern clearly indicates the seriousness with which evacuees took the warnings about possible danger.

Table 2.1
Accommodations Used by Evacuees by Distance of Residence from TMI

Distance of residence from TMI (in miles)	Accommodations				Sample Size
	Home of friend or relative	Hotel or motel	Second home or cabin	Other	
0–5	84%	11%	1%	4%	172
6–10	82%	13%	3%	2%	182
11–15	74%	20%	5%	2%	131
16–55	85%	5%	5%	5%	21

Note: Differences among groups are not statistically significant.

Accommodations

If evacuees were to stay in evacuation centers or in commercial lodgings, the financial and social impact on them would be expected to be much greater than if they stayed with relatives or friends. Table 2.1 shows the accommodations used by people living at different distances from Three Mile Island. There is little difference among the distance groups, and most persons (between 70 and 85%) went to the homes of friends or relatives.

Distances Traveled

Distances evacuees traveled is another indicator of concern about the situation at TMI. Distances, as shown in Table 2.2, were often considerable, with half of evacuees traveling more than ninety miles. It is interesting to note that the persons who lived at the greatest distance from TMI were the ones who traveled farthest, while those living close by tended to travel lesser distances. One possible explanation is that persons distant from TMI could justify evacuating only if they traveled great distances. It is also likely that any person living fifty miles from TMI who evacuated was very concerned about the situation in general and, hence, once in flight traveled a considerable distance.

Table 2.2
Distances Traveled by Evacuees by Distance of Residence from TMI

Distance of residence from TMI (in miles)	Distances Traveled			Sample size
	0–45 miles	46–90 miles	Greater than 90 miles	
0–5	33%	25%	42%	172
6–10	24%	24%	52%	182
11–15	19%	26%	55%	131
16–55	14%	14%	71%	21

Note: Differences among groups are statistically significant.

Table 2.3
Reasons for Leaving by Distance of Residence from TMI*

Distance from TMI (in miles)	Danger	Information confusing	Protect children	Protect a pregnancy	Avoid forced evacuation	Pressure from outside family	Trip planned before	Sample size
0–5	92%	78%	57%	8%	70%	23%	5%	172
6–10	92%	88%	62%	5%	79%	25%	6%	182
11–15	91%	81%	56%	10%	80%	31%	7%	131
16–55	90%	85%	60%	5%	85%	30%	5%	21

Note: Responses at different distances were not significantly different for any one of the seven reasons.

*Respondents could give more than one reason, and therefore percentages total more than 100.

Reasons for Evacuating or Staying

The reasons people gave to explain why they stayed or evacuated can provide clues as to the issues that most influenced this behavior. Respondents to the NRC survey who had evacuated were read seven possible reasons for leaving and asked if any applied to them. These reasons were obtained from pilot interviews conducted in developing the PDH survey (Houts et al., 1980). Results, shown in Table 2.3, indicate that most evacuees left because they felt there was a danger, that information was confusing, and that they wanted to avoid forced evacuation. In addition, over half evacuated to protect children. These reasons indicate serious concerns on the part of evacuees. Two reasons for leaving that could indicate less serious concerns include pressure from outside the family and a trip planned before the accident. These reasons were cited by only a minority of respondents. Interestingly, evacuees living at different distances from TMI cited the same types of reasons for leaving.

Respondents who stayed were read a list of possible reasons for not leaving, also derived from pilot interviews. The most frequent reasons cited were that the situation was in God's hands and that the respondent was waiting for an evacuation order (see Table 2.4). While these responses do not suggest high levels of concern, it should also be noted that many nonevacuees did feel in some danger since fewer than half of the nonevacuees said that they saw no danger.

There are several additional observations to be made concerning Table 2.4. The first is that three of the reasons for staying were cited more at some distances from TMI than at other distances. A larger percentage of respondents at farther distances cited "no danger," which is not surprising in view of the fact that public officials had stated that danger was greatest close to the facility. A larger percentage of respondents near TMI said they could not evacuate because they were unable to leave their jobs. This probably reflects the fact that most of those who could leave their jobs did evacuate. Finally, there was

Table 2.4

*Reasons for Staying by Distance of Residence from TMI**

Distance from TMI (in miles)	Saw no danger**	Unable to leave job**	Had no transportation	Had things to do at home	Had no place to go	Waiting for evacuation order**	Fear of looters**	In God's hands	Too sick or disabled	Sample size
0–5	38%	25%	2%	24%	12%	65%	31%	62%	2%	101
6–10	34%	21%	3%	25%	12%	64%	31%	66%	3%	202
11–15	36%	27%	5%	21%	9%	75%	20%	64%	4%	283
16–55	57%	15%	2%	24%	9%	44%	10%	64%	3%	397

*Respondents could give more than one reason, and therefore percentages total more than 100.
**Responses at different distances are significantly different.

greater concern about looters among respondents near TMI, probably because of the large number of people who had evacuated and because radio and television discussed this as a potential problem. In fact, there was almost no looting during the crisis.

A second observation about Table 2.4 is that only a small percentage of respondents cited "no transportation" or "too sick to travel" as reasons for not evacuating. This finding is of special interest to those who must plan for possible future evacuations since it indicates that only a small percent of the population would need special assistance. On the other hand, the numbers of such persons can be large in an absolute sense. For example, only 2% of respondents who did not evacuate within five miles of TMI said they were too sick or disabled to travel. This represents 0.8% of the total population. Assuming that approximately 25,000 adults live in this area, this would mean that 200 adults would need special assistance. Assuming that 400,000 persons lived within fifteen miles of TMI and that 0.8% of them were too sick or disabled to travel, this would indicate that 3,200 people would require special assistance. These figures suggest that evacuating sick or disabled people could pose significant logistical problems in planning evacuation procedures for other nuclear accidents.

Respondents in the NRC survey were also asked if there was disagreement among household members about whether to evacuate. Eighteen percent of nonevacuating households reported that there was somewhat or strong disagreement. This indicates that almost one in five nonevacuating households contained some persons who wanted to leave.

Characteristics of Leavers and Stayers

A number of analyses were carried out to identify which subgroups of the population were most likely to evacuate. We first describe the social and economic differences between evacuees and nonevacuees as determined in both the PDH and NRC July 1979 surveys. Next we use multiple regression and logit analyses to test which of these characteristics have statistically significant associations with evacuation when other factors are controlled.

PDH Survey

Table 2.5 shows that a number of characteristics of respondents within the five-mile radius of TMI were associated with evacuation. These characteristics cluster into three relatively distinct groups. The first set of variables is similar to the criteria for recommended evacuation: having a pregnant woman in the

24

Table 2.5
Selected Sociodemographic Characteristics of Households
within a Five-Mile Radius of TMI: PDH Data

Variable	Total sample	Evacuated households	Nonevacuated households
Size of household (no. of persons)	3.1	3.4	2.6
Pregnant woman in household (%)	4.5	6.8	1.0
Mean education of household head (years)	11.8	12.8	10.4
Occupation (%)			
White collar	18.1	17.0	19.6
Blue collar	57.7	60.7	54.3
Retired	14.1	12.9	15.8
Other	10.1	9.4	10.3
Family income before taxes (%)			
Less than $10,000	20.2	15.7	26.7
$10,000–$19,999	41.8	44.4	37.9
$20,000–$29,999	23.8	25.4	22.0
$30,000 and above	7.9	8.9	6.5
No answer	6.3	5.8	6.9
Male as head of household (%)	83.9	85.1	81.2
Mean age of head of household (years)	42.4	39.1	46.7
Marital status (%)			
Married	74.3	77.3	66.8
Divorced, separated, or widowed	14.7	13.3	18.7
Single	11.0	9.4	14.5
Ethnic origin (%)			
Pennsylvania Dutch	41.8	38.6	46.6
Church-going (%)	66.1	67.4	61.7
Length of stay in the area (%)			
Less than one year	6.6	7.9	4.6
1–5 years	19.4	22.0	15.2
6–10 years	15.6	18.8	10.8
Above 10 years	58.4	51.3	69.4
Left the area during two-week TMI crisis (% of evacuating households)	60.0	100.0	0.0
Sample size	691	414	277

household, having a child less than six years old in the household, being married, having a large household, and being young. The last three characteristics in this group, while not included in the governor's evacuation advisory, are closely associated with being pregnant and/or having young children. The second group indicates involvement in the community and includes length of residence, owning a home, being Pennsylvania Dutch, and church attendance. The third group reflects socioeconomic status and includes education, income, and occupation. The results within the first group are consistent. All traits associated with the governor's advisory occur with higher frequency among evacuees than among those who stayed. The community involvement group of variables is not as consistent. While longer residence, home ownership, and being Pennsylvania Dutch are associated with staying, church attendance is higher among evacuees. Among the socioeconomic indicators, results are again consistent, showing people with higher education, higher income, and white-collar occupations more likely to evacuate.

One feature of this population that deserves further explanation is the relatively large number of persons within five miles of TMI who identify themselves as "Pennsylvania Dutch." As was pointed out in the Introduction, respondents identifying themselves as "Pennsylvania Dutch" are descendants of German and Swiss immigrants who settled in this area during the late seventeenth and early eighteenth centuries. Only a small percent of these people are Amish or Mennonites, subgroups of the Pennsylvania Dutch, who, in the public's mind, often represent the group. The fact that there are more "Pennsylvania Dutch" among nonevacuated households than among evacuated households (46.6% vs. 38.6%), may reflect a greater feeling of having "roots" in the area and possibly a more conservative outlook as well.

NRC Survey

Table 2.6 shows demographic characteristics, gathered from the NRC survey, of evacuated and nonevacuated households at different distances from TMI. The findings for the five-mile ring in Table 2.6 were collected from the same geographic area using the same sampling techniques as for the PDH data presented in Table 2.5. Therefore, the degree of agreement is an indication of the reliability of data obtained through these telephone surveys. It can be seen that, when the two surveys asked the same questions, most of the results are similar. A noticeable difference between the two surveys' findings is in the distribution of occupations. This may be because questions about occupation were phrased somewhat differently in the two surveys. The percent of respondents who own their homes is slightly larger in the PDH sample (78% vs. 71%). The relative distribution of incomes among evacuees and nonevacuees is similar in the two samples except that the proportion of respondents in the highest income category (over $30,000) is higher among evacuees in the PDH sample and higher among nonevacuees in the NRC sample. The estimated percent of households that evacuated is quite similar (60% in the PDH survey and 63% in the NRC survey).

An examination of Table 2.6 shows that differences between evacuated and nonevacuated households are generally the same at all distances from TMI. One exception is that the percent of persons with incomes over $30,000 who stayed is greater close to TMI, but less beyond ten miles. However, the fact that evacuees have higher incomes for fourteen out of sixteen comparisons

26

Table 2.6

Selected Sociodemographic Characteristics of Households at Various Distances (Miles) from TMI: NRC Data

Variable	Evacuated households				Nonevacuated households			
	0–5	6–10	11–15	16+	0–5	6–10	11–15	16+
Size of household (no. of persons)	3.18	3.30	3.35	3.33	2.49	2.96	2.69	3.23
Pregnant woman in household (%)	3.5	3.8	6.9	9.5	5.9	4.5	2.5	5.3
Mean education of household head (years)	12.39	13.01	14.00	13.47	11.84	12.3	12.4	12.18
Occupation (%)								
White collar	31.0	39.7	59.5	47.6	29.6	37.6	33.5	30.9
Blue collar	49.9	34.1	22.2	28.6	31.6	31.4	34.2	36.5
Other	28.1	26.2	18.3	23.8	38.8	31.0	32.3	32.6
Family income (%)								
Less than $10,000	22.8	15.0	14.5	10.5	32.7	17.4	25.4	25.2
$10,000–$19,999	49.4	41.0	37.1	15.8	38.8	46.2	41.2	44.0
$20,000–$29,999	21.6	31.2	28.2	57.9	16.3	20.0	22.7	19.9
$30,000 and above	6.2	12.7	20.2	15.8	12.2	16.4	10.8	10.8
Male as head of household (%)	79.7	84.6	85.5	95.2	77.2	81.2	79.2	82.4
Mean age of head of household (years)	39.60	41.7	40.6	38.57	48.9	49.6	47.9	46.34
Marital status (%)								
Married	71.5	83.4	77.9	95.2	62.0	71.3	66.1	76.5
Divorced, separated, or widowed	14.5	12.2	13.0	4.8	20.0	16.3	18.7	17.0
Single	14.0	4.4	9.1	0.0	18.0	12.4	15.2	6.5
Evacuation within the ring (% of households)	63	49	32	5	—	—	—	—
Sample size	172	182	131	21	101	202	283	397

suggests that this reversal may not be significant. Another exception is in the occupation category, where the percent of blue-collar respondents is greater among evacuees within ten miles, but less among households beyond ten miles. It is difficult to interpret this finding, especially in view of the large percent of respondents assigned to "other" in the NRC survey data.

The differences between evacuees and nonevacuees in Table 2.6 are generally the same as in Table 2.5. Variables that indicate similarity to the group that was advised to evacuate are consistently associated with evacuation. Socioeconomic status, as indicated by income and education, is also associated with evacuation. These socioeconomic relationships may reflect ease of evacuating or, possibly, more critical attitudes toward the information provided during the crisis. The third factor, community involvement, is associated with staying rather than evacuating. This association may reflect conservatism among people who have lived in the area for a long time or, possibly, greater feeling of involvement with the community (and therefore greater reluctance to leave). These are tentative hypotheses since the findings on which they are based are simply associations, and there are many factors, including interrelationships among the explanatory variables, that can play a role. In order to explore these hypotheses further, multiple regression analyses are utilized. The advantage of using multiple regression analysis is its ability to consider all relevant independent (explanatory) variables together in predicting the outcome of a dependent variable.

Multiple Regression Analyses

The dependent variable in these multiple regression analyses is dichotomous (that is, left or stayed). One of the assumptions of ordinary least squares linear regression (constant variance of the error term) is violated in such situations, and there are reasons for preferring an S-shaped, or logistic, function for modeling a dichotomous dependent variable. However, in most situations the results from linear and logistic regression are very similar (Cleary and Angel, 1984). Thus, to simplify comparisons with other chapters, linear regression results are presented here. Logistic regressions were also estimated using maximum likelihood estimation techniques, and the results showed the same variables to have statistically significant associations with evacuation as in the multiple regression analyses shown here (Hu and Slaysman, 1984).

Table 2.7 includes the multiple regression results based on both the PDH and NRC data. The first six variables, through distance from TMI, are those related to the governor's evacuation advisory. In the PDH data, having a larger household and having children under six years of age are associated with evacuation. In the NRC data three of the variables associated with the governor's advisory are significantly related to evacuation: living within fifteen miles

Table 2.7
Regression Results of Household Decision to Evacuate: PDH and NRC Data

Variable	PDH Beta Coefficients	NRC Beta Coefficients
Similarity to persons advised to evacuate		
Age of respondent	−.04	−.08**
Pregnant person in household	.03	.02
Married (compared to single, widowed, separated, or divorced)	.01	.05
Household size	.15**	.03
Child less than 6 years old in household	.22**	.13**
Distance from TMI (each group compared to group from 41–55 miles from TMI)		
0–5 miles	X	.49**
6–10 miles	X	.41**
11–15 miles	X	.28**
16–25 miles	X	.05
26–40 miles	X	.01
Community involvement		
Church attendance	.01	X
Ethnic origin (Pa. Dutch compared to all others)	−.01	X
Length of residence	−.09*	−.05
Own home vs. rent	−.02	−.04
Socioeconomic status		
Education of respondent	.11**	.06*
Income of household	.03	−.02
White collar (1) vs. other occupational groups (0)	−.03	.04
R^2	.17	.27
F ratio	10.47**	33.02**
Sample size	692	1504

*p < .05.
**p < .01.
X = variable not included in study.

of TMI, having a child less than six years old in the household, and younger age. Distance is an extremely important variable in explaining the decision to evacuate, as shown by the size of the regression coefficients. For example, households within the five-mile area are shown to have a 61% higher chance of evacuating than do residences beyond forty miles. The probability decreases to 45% for six to ten miles and 29% for eleven to forty miles. The evacuation rate for the sixteen-to-forty-mile group is not significantly different from that of residents beyond the forty-mile ring.

None of the "community involvement" variables is associated with evacuation in the PDH sample. In the NRC data, only living in the area a short time is significantly related to evacuation. Among the socioeconomic variables, more highly educated respondents are more likely to evacuate in both the PDH and NRC samples.

The results from these models are very helpful in elucidating the reasons for evacuation, but it is important to emphasize that they explain relatively little of the variance in behavior. The analysis of data from respondents living within five miles of the reactor explained less than 20% of the variance in behavior, and the model based on the NRC data, which included distance from the reactor in addition to the other explanatory variables, explained less than a third of the variance. Thus, these models leave a great deal unexplained about the evacuation behavior of residents after the accident.

Conclusions

Evacuation behavior is one indicator of the impact of a crisis, and the findings on evacuation reviewed here indicate that the TMI accident had a substantial impact on the population near the nuclear power plant. This is shown by the large number of persons who evacuated, the distances people traveled (over half traveled distances greater than ninety miles), the reasons evacuees gave for leaving (for example, the situation was dangerous and the desire to protect children), and the fact that most people left shortly after the governor's advisory. The impact was less among those who stayed, though it should also be noted that over half of the nonevacuees said they felt there was some danger. In addition, almost 20% of families that stayed contained some persons who wanted to leave.

Analyses to determine which subgroups of the population were most affected by evacuation indicate that people with characteristics similar to those listed in the governor's evacuation advisory were the most affected. Thus, households in which there were children less than six years old and which were located close to TMI were the most likely to evacuate. It is as if individuals located themselves on a continuum of similarity to the group that was advised to evacuate (families with young children or pregnant women living within five miles of TMI), and the greater the similarity, the greater the probability of evacuation. Contrary to the Perry, Lindell, and Greene (1981) findings, younger people were more likely to evacuate. However, this is consistent with studies cited by Stallings (1984), which also reported that young people are most likely to evacuate. In the TMI situation the tendency for younger people to evacuate is perhaps due to the emphasis the governor placed in his advisories on young people leaving the area. In addition, socioeconomic status, especially education, was associated with evacuation, which is also consistent with findings in other evacuations reviewed by Stallings. The greater tendency for highly

30

educated persons to evacuate may be the result of its being easier for them to leave due to greater ability to be away from their jobs. It may also reflect a skeptical attitude toward the conflicting information available about the situation at TMI. A third group of variables, indicating degree of involvement in the community, also showed some association with evacuation, though only one of these variables was significantly related to evacuation in the multiple regression analyses: how long their families had lived in the area.

In comparison with studies of other disasters, the TMI accident generated one of the largest population evacuations. Stallings (1984), in a review of the evacuation literature, noted that the median disaster evacuation involved fewer than 1,000 persons and that 94% of all evacuations involved fewer than 100,000 persons. Thus the 144,000 evacuees during the TMI crisis places this among the few largest evacuations. Furthermore, distances traveled by evacuees were much larger than usually occurs in natural disasters, with half of the evacuees traveling more than ninety miles compared to an average distance of thirteen miles reported by Stallings in his review. On the other hand, the length of evacuation following the TMI accident was relatively short, especially when compared to the thirty-one-day average evacuation reported by Melick (1976) for Hurricane Agnes.

As in the Perry, Lindell, and Greene (1981) study of natural disasters, warnings by authorities appear to have been a stimulus in evacuation decisions. However, the TMI crisis differs from the disasters they studied in that there were no visible warning signs (such as high waters) to reinforce the issued warnings. As a result, the evacuees at TMI were primarily dependent on information contained in the advisories issued by the governor to judge their degree of risk. Most TMI evacuees stayed with either relatives or friends, which is consistent with Perry, Lindell, and Greene's (1981) findings for evacuees from floods and with Stallings' (1984) review of earlier evacuation studies. This may have softened the impact of evacuation, since staying with friends or relatives provided a degree of support and also limited the expenditures for each family.

In the preface to this book we raised the question of whether reactions to the TMI crisis were similar to or different from reactions in other crisis situations. Stallings (1984) feels that evacuation behavior during the TMI crisis was similar to what would have been predicted from other disaster situations. In a review of the literature on evacuation at Three Mile Island, he concluded that "voluntary evacuation at Three Mile Island did not differ significantly from those taking place in natural disasters. Therefore, no special plans, policies or procedures seem needed over and above those in place for other kinds of disaster evacuations." Our conclusions, based on the findings presented in this chapter, are somewhat different. We agree with Stallings that several aspects of the TMI situation are similar to those reported for natural disasters. These include the facts that evacuations were largely in family groups, that persons of higher socioeconomic status were more likely to evacuate than persons of lower socioeconomic status, and that most evacuees stayed with family or friends. However, there also were important differences. In the TMI crisis, evacuees traveled greater distances but stayed away for shorter times than in most natural disasters, and the number of people who evacuated at TMI was much larger than is usually the case for natural disasters. The large evacuation near Three Mile Island occurred in part because most people perceived radiation as spreading through the air and therefore affecting a large geographic area. Natural disasters, on the other hand, usually affect only limited geographic areas.

Another important difference between the evacuation at TMI and natural disaster evacuations is in the role of disagreements among experts as a reason for either leaving or staying. Stallings (1984) noted that conflicting information is often cited as a reason for not evacuating in natural disasters. However, both the NRC and PDH July surveys found that approximately 80% of those who evacuated at TMI said that confusing information was one of the reasons they decided to leave. This is in contrast to only 40% of nonevacuees who cited conflicting reports as a reason for staying (Brunn et al., 1979). This suggests that confusing or contradictory reports about danger may play a different role in technological as compared to natural disasters. In natural disasters the source of the danger is something that people experience or hear about regularly, such as the weather or a river, and, as a result, many people feel that they are capable of judging the danger themselves. Even if they have not had personal experience with a certain kind of natural disaster, they probably have read or heard about similar situations. Therefore, when experts disagree as to the degree of danger, people feel able to make their own judgments about the degree of risk they face. In contrast, people living near Three Mile Island did not feel capable of judging the degree of danger because the source of danger was unfamiliar and not easily understood. Therefore, when the experts disagreed, people felt they could not evaluate the danger themselves, which increased their concerns. It is likely that the association between uncertainty and evacuation found at TMI will be repeated when the crises involve complex technologies that are not easily understood by the general public.

A third important difference is that in most natural disasters the number of people evacuating is smaller than the number of people advised to leave, while, during the TMI crisis, the opposite occurred. The TMI evacuation advisory was directed at a relatively small number of people, that is, pregnant women and families with preschool children living within five miles of Three Mile Island. Yet the number who evacuated was many times that number. Interestingly, analyses in this chapter and in Houts et al. (1984) indicate that people judged their levels of risk based on their similarity to the group advised to evacuate. This suggests that people living near TMI did pay attention to what was said in the advisory. However, one part of the advisory that many people did not accept was that danger was limited to that subgroup. This suggests that public officials dealing with similar crises can expect that evacuation advisories directed at subgroups of the population will affect the behaviors of many people outside those subgroups.

In conclusion, several findings are different from what is known about responses to most natural disasters. These can have important implications for public officials responsible for planning and directing evacuations in both man-made and naturally caused disasters. They suggest that, when the source of danger is airborne or, like radiation, can travel through the air, the scope of the evacuation can be expected to be much greater than is usual for most natural disasters. Second, they suggest that, in situations where the danger is the result of complex, unfamiliar technologies, lack of consistency among experts about the degree of danger will have the effect of increasing rather than decreasing the size of the evacuations. And third, they suggest that advisories directed at selected subgroups of the population are likely to affect the behaviors of many more persons than just members of those subgroups, and that people will judge their risks in terms of their similarities to the subgroup to which the advisory was directed.

3

Economic Impact on Persons and Businesses in the Vicinity of Three Mile Island

Overview

Short-term economic losses as a result of the TMI accident were much smaller than for natural disasters such as Hurricane Agnes, the Tug Fork valley flood, and the Mount Saint Helens volcanic eruption. Of the short-term effects, the loss to business, totaling around $81.9 million, constituted the greatest impact. The evacuation costs to households reached only $5.99 million. Health-related costs were the smallest expenditure resulting from the TMI crisis. Long-term economic effects were not large for persons living in the immediate vicinity of the plant, and there was no indication of any significant change in property values near TMI due to the accident.

Most of the long-term economic costs, which were estimated to exceed a billion dollars by 1988, were borne by insurance companies, the federal and state governments, other electric power companies, and purchasers of power throughout the United States. Some of these funds were clearly identifiable, as in the case of insurance payments and government appropriations. Others, however, involved complex accounting and legal arrangements among companies and between companies and governmental regulating agencies. What can be concluded is that there was cooperation among industry and governmental groups to minimize the economic impact on customers of General Public Utilities (GPU), which owned Three Mile Island, and also to insure the company's economic survival. The stockholders of GPU sustained costs in the form of lost dividends (estimated at more than $800 million from 1979 through 1987). A short-term drop in the value of the company's stock followed the accident. However, the stock price gradually rose to where, in 1987, it was worth 50% more than in 1979, which is still less than the 100% average increase for all other utilities serving eastern states during that period.

Worldwide there was a cost in the decreased performance of pressurized water reactors that was estimated to equal the cleanup costs. There was also a loss in stock values, at least through 1980, for all utility companies that had invested heavily in nuclear power and especially those that used Babcock and Wilcox designs (the same design used at Three Mile Island). It is not known whether the value of these stocks rebounded after 1980 as GPU did. Costs borne by society as a whole include the economic effects of more stringent monitoring of nuclear power plants by the Nuclear Regulatory Commission following the TMI accident and the impact of the accident on the building of new nuclear power plants in the United States. Among the unmeasurable economic benefits for society is the decreased likelihood of future accidents at nuclear power plants as a result of what was learned about the causes of the TMI accident and more stringent monitoring by the Nuclear Regulatory Commission.

Background

All disasters have both short- and long-term economic consequences. Short-term economic impacts are measured in terms of the costs incurred immediately as a result of the accident, and their magnitude is a function of the size of the affected population. The short-term economic impacts of the TMI accident can be divided into those affecting individuals and those affecting businesses. These impacts include health-related costs and costs incurred due to evacuation. Long-term economic impacts of the TMI accident, affecting both the private and public sectors, include costs of cleanup of the damaged reactor, changes in real estate values, and increases in electricity rates. Many of these long-term economic impacts are difficult to measure, although they potentially have greater cost implications than the short-term economic impacts.

Review of Costs of Recent Disasters

Following are some recent examples of economic damages caused by natural disasters. Hurricane Agnes, in June 1972, caused flooding in the Wilkes-Barre area that resulted in $3.4 billion worth of damages, including the destruction of 25,000 homes and 2,728 business establishments (Melick, 1976). A study of flood trauma for the 1977 flood in New York's Tug Fork valley (Allee, 1980) estimates business losses at $44.9 million, emergency costs of evacuation at $25.8 million, and physical damage at $126.6 million. The Mount Saint Helens volcanic eruption in 1980 destroyed three billion board feet of timber valued at approximately $400 million, with total damages in property and crops estimated at more than $1.8 billion in the vicinity of the volcano and those areas that suffered from ash fall (U.S. Senate Hearings, 1980). Wright and colleagues (1979) estimated the effects of the major floods, tornadoes, and hurricanes that occurred between 1960 and 1970 in the United States and found no discernible effects of any of these types of disasters on changes, at the county or census tract level, in population or available housing in the areas

35

affected. They also examined other measures of county level impact, such as housing values, rents, age composition, education level of the population, and family income, and found no consistent patterns. Thus, they conclude that there were no direct, long-lasting economic effects on the county level.

The Nuclear Regulatory Commission made an estimate of the cost of evacuation during the TMI crisis (Flynn, 1979). In that study, median values of $100 per evacuated household and $100 per evacuee who lost pay were used to estimate total cost. The resultant estimates were about $9.8 million for total costs of evacuation, $9.2 million for income loss, and $0.4 million for other expenses incurred by nonevacuees, totaling $19.4 million excluding the insurance compensation, which amounted to $1.2 million. Thus, the NRC study concluded that the total cost during the TMI accident, minus insurance compensation, was about $18 million.

Scope of the Economic Impact

Short-Term Impacts

Economic impacts that occur during and immediately after a crisis are designated as short-term economic impacts. We examine separately the two types of short-term costs: those affecting the individual and those affecting businesses. The impact on individuals is further subdivided into the two primary causes of this impact: evacuation costs and health-related costs. All of the cost data discussed in this section were obtained from the PDH and NRC surveys in July 1979 and from a survey conducted by the Pennsylvania Department of Commerce, reported by the Commonwealth of Pennsylvania's Governor's Office (1979).

On Individuals: Evacuation Costs

The household costs of evacuation can be divided into two categories: (1) costs involving direct outlays relating to evacuation activities—for example, travel (automobile, bus, train, or airplane), lodging, meals, and other incidental costs; and (2) costs resulting from evacuation—for example, the loss of take-home pay (for wage earners). Some of these evacuation costs may be reimbursed by insurance companies, so that the net loss to households may be less than their initial payment. But since insurance reimbursement, from the societal point of view, is a form of transfer payment, the total social costs of evacuation should be the net resources used (or lost) due to evacuation, regardless of which party paid the costs (as long as there is no double counting).

Table 3.1
Median Cost of Evacuation during TMI Accident: NRC Data

Costs	Distance (miles)		
	0–5	6–10	11–15
Evacuation costs ($)	150	100	100
Percent of evacuated families that lost pay	32	23	17
Work days lost	4.2	4.2	3.3
Pay loss ($)	116	120	130
Percent of evacuated families that had other expenses	95	6	6
Other expenditures ($)	79	52	50

To estimate the overall economic costs of evacuation in the community, it is necessary to first estimate the total number of households and population evacuated during the two-week period of the accident. Based on information on evacuation behavior presented in Chapter 2, we estimate that about 50,000 households or 144,000 individuals evacuated from the area within fifteen miles of Three Mile Island during the crisis. The data also indicate that, on the average, the number of days away from the area was between four and five.

The costs of evacuation include household expenditures on food, transportation, and lodging. Reported costs range from a few dollars to $2,000. Since the distribution of evacuation costs is rather skewed, the median value is used for the cost estimation. The median costs of evacuation were $150 within a five-mile radius and $100 beyond a five-mile radius. The mean costs of evacuation are about $50 higher than the median values. The reason that the costs are relatively lower than might have been expected is that two-thirds of the evacuated households went to stay with relatives, 15% went to stay with friends, and only 8% went to a hotel or motel. Other expenditures, including incidental costs incurred as a result of evacuation (such as clothing, gifts for friends, telephone calls, etc.), ranged from $50 to $79 for some evacuated households.

In addition to the direct evacuation costs, 32% of the households within a five-mile radius of the plant lost pay. This compares to 23% and 17% for households in the six-to-ten- and eleven-to-fifteen-mile rings, respectively. The average work days lost ranged from three to four. Median overall pay loss ranged from $116 to $130 for the three distance areas covered in the survey. Table 3.1 provides a summary of the median values of evacuation costs and percent of households that incurred a pay loss.

Based on these median estimates, the percent of evacuation, and the number of households and population in these areas (shown in Table 3.2), extrapolated cost estimates are presented in Table 3.3. The household costs of evacuation

Table 3.2

Household Costs of Evacuation during the Two-Week Period of TMI Accident,
Estimates Based on Median Value: NRC Data

Costs (in dollars)	Distance (in miles)			Total by cost category
	0–5	6–10	11–15	
Evacuation costs	1,349,000	1,970,000	2,301,000	5,690,000
Pay loss	337,000	545,000	497,000	1,379,000
Other expenditures	67,000	57,000	69,000	193,000
Total by distance	1,753,000	2,572,000	2,867,000	7,192,000

are about $7.2 million for the entire fifteen-mile ring. These cost estimates are much smaller than the NRC estimates of about $18 million. The difference is attributed to smaller pay loss and other incidental cost estimates in this study than in the NRC study (Flynn, 1979). The Flynn study estimates that 56% of the evacuees within the fifteen-mile area who lost work also lost pay. This study (Hu and Slaysman, 1984) analyzed the responses to the NRC survey in more detail and developed separate estimates for different distances within fifteen miles. These estimates, which are explained in detail in Hu and Slaysman (1984), are as follows: only 32% of workers who lost work also lost pay within five miles; 23%, between five and ten miles; and 17%, between eleven and fifteen miles. As a result, we estimate that only $1.5 million was lost in pay within the fifteen-mile radius where Flynn estimates a $9.5 million loss.

It should be noted that a large portion of those people who had a work loss did not incur loss of pay. For instance, about 40% of workers within five miles of TMI who lost work days did not lose any pay. This compares to 60% of workers who lost work days in the six-to-ten-mile area and 81% in the eleven-to-fifteen-mile area. Although these workers did not lose pay, the business or the industry sector must have absorbed the loss of production. This is figured into the total loss suffered by the business or industry sector. Since it is already accounted for, it should not be counted again.

Although one may argue that additional food expenditures by the relatives or friends of evacuated households should be considered in estimating the total cost to society, it was decided not to include such expenditures in this study. From the societal point of view, food costs, like insurance, represent a transfer payment between two parties. Had there been no TMI accident, the evacuated households would have had to spend money on their food items. Thus, no additions should be made to the cost calculation. (The NRC study did include food costs, which inflated their estimates.) Similarly, insurance for the TMI facility paid an average of $88 per household, or about $1.2 million ($643,000 for the 0–5 mile area, $424,000 for the 6–10 mile area, and $148,000 for the 11–15 mile area). Since these are also transfer payments between two parties, one should not add or subtract them in estimating the total social costs of

Table 3.3
Changes in Health-Related Behavior Due to TMI Crisis: 0–5 Mile Ring

	Apr. 1979 (First 2 weeks after accident)		July 1979–Jan. 1980	
	Quantity*	Costs($)**	Quantity*	Costs($)**
Physician visits	290	4,350	235	3,525
Work days lost	8,870	266,100	8,597	257,100
Sleeping pills (tablets)	298	36	153	18
Tranquilizers (tablets)	802	96	171	12
Alcohol (servings)	506	336	—	—
Cigarettes (packs)	1,900	950	1,753	876
Total		271,868		261,531

*The quantities are derived from regression coefficients representing the impact of stress on each type of cost. See Hu and Slaysman (1984) for an explanation of these analyses.

**The unit costs were $15 for each physician visit, $30 for each work day lost, $.12 for each sleeping pill or tranquilizer, $.70 for one alcoholic beverage serving, and $.50 for one pack of cigarettes.

evacuation during the TMI accident. However, from the households' point of view, these insurance payments did reduce their costs of evacuation.

In summary, the estimates of the evacuation's economic impact presented here are based on more detailed analysis of the data than in the Flynn report as well as more conservative and, we feel, more reasonable assumptions. As a result, our estimate of $7.2 million costs is smaller than the $18 million estimated by Flynn.

On Individuals: Health-Related Costs

In this section the economic impact of health-related behaviors due to the TMI crisis (for example, physician visits, work days lost, and use of sleeping pills, tranquilizers, alcohol, and cigarettes) is discussed. A detailed presentation of the statistical analyses involved in making these estimates can be found in Hu and Slaysman (1984). In computing these estimates, we calculated what we considered to be maximum costs. Therefore, when multiple interpretations of the data were possible (which often occurs with retrospective reporting), the interpretation that resulted in the maximum impact was accepted.

Data concerning health-related costs are available in the PDH July 1979 survey, in which respondents were asked questions about physician visits, work days lost, and the use of sleeping pills, tranquilizers, alcohol, and cigarettes during and after the crisis period. The first step in the analysis is to assign dollar costs to these behaviors. Next, the percent of these costs due to the TMI crisis is estimated by determining the degree of relationship between these behaviors and reported stress during the crisis utilizing multiple regression analyses. The results of these analyses are shown in Table 3.3.

Table 3.3 indicates that the largest health-related cost is due to work days lost, and the next largest is due to the increase in physician visits. The costs of increased consumption of sleeping pills, tranquilizers, alcohol, and cigarettes are minimal. Among them, cigarettes account for the largest share of the costs. These analyses indicate that the maximum health-related costs during the two weeks following the accident were about a quarter of a million dollars, with another quarter of a million expended through January of 1980. Thus the total health-related costs within the five-mile radius of TMI are estimated at $.5 million.

On Businesses

The Pennsylvania Department of Commerce conducted a survey of business and manufacturing industry within a twenty-mile radius of the TMI plant to determine the accident's immediate and short-term economic impact (Commonwealth of Pennsylvania, 1979). The survey shows that most of the economic impact from TMI on the local industry was the loss of production during the first week of the accident—March 30 to April 6, 1979—as a result of the evacuation. During the first week, the manufacturing firms lost an estimated $7.67 million in value of production, or $6,725 average for 1,141 establishments in the area. The losses in value of production were caused by plants closing and by disruptions in the supply of raw materials and in the marketing of finished goods and products to consumers. Most of these disruptions resulted from the voluntary evacuation and concern over the possible radiation on processed food products.

The Pennsylvania Department of Commerce study also found that over the entire one-year period only 20% of the 1,141 manufacturing establishments experienced any negative fiscal impact related to the TMI accident. Two-thirds felt no fiscal impact at all. Year-long losses were claimed by only 7% of the establishments affected by the accident.

The accident also had a direct impact on nonmanufacturing businesses. The nonmanufacturing sector lost an estimated $74.2 million in value of production. The absolute losses of the nonmanufacturing sector were greater than those of the manufacturing sector. However, since there were 20,197 nonmanufacturing firms in the area, the average loss per establishment during the first week was $3,673, which is less than the $6,725 average loss of the manufacturing plants. Over the entire year, 28% of the more than twenty thousand establishments experienced negative fiscal impact from the accident.

The tourist industry in the TMI area estimated its loss at about $5 million during the first month after the accident. However, because of the publicity surrounding the accident, the tourist business recovered and in fact surpassed the level it enjoyed during the preaccident period.

Two other minor economic impacts were the additional public sector personnel expenses and the corresponding increase in administrative costs during the TMI accident. On these, the state and local governments expended more than $900,000.

Long-Term Impacts

In the Immediate Vicinity of TMI
There has been considerable discussion about the effects of the Three Mile Island crisis on the value of nearby property. Two studies (Gamble and Downing, 1981; Nelson, 1981) have examined sales of single family homes before and after the March 1979 accident, within and beyond the TMI area. Both studies concluded that the accident had no measurable effects, positive or negative, on the value of single family residential properties within a twenty-five-mile radius of the plant, irrespective of the original value of the properties. This finding was further supported by Goldhaber et al.'s (1983) study of geographic mobility (changing residences) following the crisis, which found no evidence that the crisis had affected mobility within five miles of TMI.

Another possible long-term impact on people living near Three Mile Island was increased rates for electricity. Approximately a third of the geographic area within fifteen miles of TMI was served by General Public Utilities (GPU), the owner of Three Mile Island. (Metropolitan Edison, the operator of the TMI facility at the time of the accident, is a subsidiary of GPU.) While GPU rates did rise following the crisis, the increases were approximately the same as those of other Pennsylvania utilities during the period. The magnitude of these rate changes is discussed in more detail below.

On Society as a Whole
The overall long-term costs of the nuclear accident at Three Mile Island have been substantial. General Public Utilities estimated the total cost of the clean-up, including the costs of maintaining the facility while it was not producing power, at a billion dollars. In addition, there were costs of purchasing replacement power at prices above what it would have cost had TMI been operational. According to reports issued by GPU, as of July 31, 1987, 41% of the costs of the cleanup had been paid by insurance, 35% by GPU and its customers, and 24% by governmental and industry groups, including the U.S. Department of Energy, the states of Pennsylvania and New Jersey, and utility groups both in the United States and Japan. The amounts contributed by insurance and governmental and industry groups are shown in Table 3.4. The contributions of GPU stockholders and customers are not included in Table 3.4 because they are difficult to verify due to the complex accounting procedures involved. We have, therefore, taken a different approach to assessing the costs borne by these groups, comparing their experiences with what happened with owners and customers of other utilities.

Stockholders were vulnerable to two types of losses: loss of dividends and loss in the value of stock. The loss of dividends is an economic cost since it represents reduced production and earnings of the company. Loss in the value of stock, on the other hand, is not considered an economic loss since stock sales are simply payment transfers, which do not represent any change in the

Table 3.4
Insurance, Government, and Industry Contributions
to Cleaning up the Damaged Reactor at Three Mile Island*

Sources of funds	Contributions as of 7/31/87 (thousands of dollars)
Insurance	305,990
U.S. Dept. of Energy	72,043
State of Pennsylvania	25,004
State of New Jersey	7,554
U.S. nuclear utilities	65,028
Electric Power Research Institute	1,828
Japanese nuclear power industry	9,530
Total	486,977

*Excluding GPU contributions.
Source: General Public Utilities Corporation.

production of goods and services. The average dividend paid to GPU stockholders from 1974 through 1978 was $1.70 a share. In 1979, the year of the accident, dividends dropped to $1.20 a share and no dividends were paid until 1987, when an estimated $.45 a share will be paid (Value Line, Inc., 1987). If the 61.25 million shares of common stock outstanding at the time of the accident had been paid dividends of $1.70 per share from 1979 through 1987, more than $937 million would have been paid to stockholders rather than the approximately $101 million that was paid during that period. This suggests a loss to stockholders of more than $800 million. Furthermore, this estimate is conservative because, during that period, the other thirty-four utilities that serve the eastern United States increased their dividends by an average of 36% (Value Line, Inc., 1987).

The long-term value of the company's common stock was also affected by the TMI accident, and while this is not an economic loss for society, it could be so for individual stockholders who sold their holdings following the accident. The value of GPU stock dropped from approximately $18.00 a share just prior to the accident to a low of $3.38, but it subsequently rose, especially after the restart of TMI's undamaged reactor in 1985, to more than $27.00 a share in October 1987. This increase of 50% over the eight-year period is half of the 100% average increase in the value of the other thirty-four electric generating utilities in the eastern United States during that time (Value Line, Inc., 1987). Had GPU stockholders experienced a 100% instead of a 50% increase in the value of their stock, their equity would have increased by an additional $551 million. It should be noted, though, that the value of the stock may eventually recover fully, but only for those who continue to hold their GPU stock.

In order to estimate the effects of the TMI accident on GPU's four million customers, we compared the residential rates paid by GPU customers in

FIGURE 3.1

RESIDENTIAL ELECTRICITY RATES
ALL PENNSYLVANIA UTILITIES 1975 - 1986

(SOURCE: PA PUBLIC UTILITY COMMISSION)

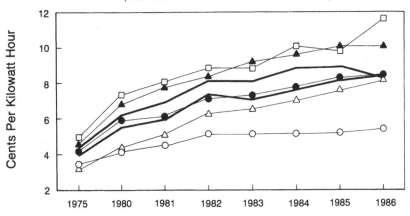

Heavy lines = subsidiaries of General Public Utilities, owners of Three Mile Island

Pennsylvania before and after the accident with those of other utilities in Pennsylvania. The results, in Figure 3.1, show that GPU residential rates were in the middle of the range of rates charged by Pennsylvania utilities from before the accident, in 1975, through 1986. Therefore, while it is possible that the rates would have been lower had TMI been functioning, it is clear that GPU customers did not experience hardship relative to other residential ratepayers in Pennsylvania. Rates paid by GPU customers in New Jersey also stayed within the range charged by other utilities in the state.

Had the customers and stockholders of GPU borne the full costs of the accident, the results would have been staggering for both. The financial impact of the accident was softened by the contributions of government and industry to the cleanup (totaling more than $180 million by July 31, 1987). In addition to these direct contributions, both groups took additional actions to protect GPU customers and stockholders from the full impact of the accident. For example, much of the replacement power that GPU purchased to take the place of power that TMI would have generated was repriced so that GPU payed a lower price than would normally have been the case. The state of Pennsylvania also contributed to the lowering of replacement power costs by exempting GPU's purchases of replacement power from a state revenue tax. The Pennsylvania Public Utilities Commission helped to keep the company viable by allowing GPU to amortize the TMI plant at an accelerated rate, thereby improving the company's cash flow when funds were needed for the TMI cleanup operations.

The economic impact of the TMI accident internationally was studied by Evans (1982), who found a significant drop in the performance of pressurized water reactors (the type involved in the TMI accident) throughout the Western world following the accident. This was presumably due to the reducing of station output and shutting down of reactors for inspection because of concern about the safety of that type of reactor system. This effect was specific to pressurized water reactors and did not extend to other types of reactors. Evans estimated the cost worldwide to be of the same order as the TMI cleanup operation.

Another potential long-term impact was the effect on the value of stocks of utilities that have substantial investments in nuclear power. Bowen et al. (1983) studied the stock values before and after the TMI accident of four groups of utilities: 1) those with a major commitment to nuclear energy (i.e., at least 20% of the generating capacity was nuclear); 2) those with plants built by Babcock and Wilcox, the contractor for the TMI facility; 3) those with between 10 and 20% nuclear capacity; and 4) those with no present or planned nuclear capacity. They found a significant downward shift in the prices of the companies' stocks as a function of their commitment to nuclear power and/or of having Babcock and Wilcox designed plants. The study also showed that the price decline persisted at least through the beginning of 1980, one year after the accident. No studies have been published on whether these effects on the value of stocks persisted after 1980.

Another important, but difficult to estimate, long-term cost of the TMI accident is the expenses borne by nuclear power companies because of more stringent monitoring and more rigorous design requirements for nuclear plants by the Nuclear Regulatory Commission. The other side of these costs is the considerable potential savings because of the reduced likelihood of future accidents at nuclear power plants.

Conclusions

Short Term

Short-term costs of the TMI accident were largely a result of the extensive evacuation that occurred during the crisis period. Because there was no physical damage to the surrounding area, these costs were generally smaller than those reported for most natural disasters. Table 3.5 summarizes the major types of short-term costs of the accident for businesses and persons located in the vicinity of Three Mile Island. While the types of costs were computed for different geographic areas, the differences between categories are sufficiently

Table 3.5
Summary of Economic Costs of the TMI Accident on Businesses and Persons near Three Mile Island (millions of dollars)

Short-term impacts, Individual Costs:	
Costs of evacuation (within 15 miles)	
Net costs incurred to households	5.99
Costs reimbursed by insurance company	1.20
Total evacuation costs	7.19
Costs of health-related behavior (within 5 miles)	.50
Business Costs*: (within 20 miles)	
Manufacturing industries	7.67
Nonmanufacturing industries	74.20
Total business costs	81.87

*This total does not include tourist and government costs.

large that certain patterns can be discerned: that the largest short-term costs were to businesses in the area and that most of the evacuation costs to households were borne by the household members, with less than 20% of those costs reimbursed by the company's insurance. Health-related costs were not large in comparison to the other short-term costs. Furthermore, as will be seen in Chapter 4, Blue Cross/Blue Shield claims for the area near TMI showed little change following the accident, which is an additional indication that the health-related costs of the accident were relatively small.

Long Term

There is little evidence of significant long-term costs to people living in the vicinity of the reactor. Real estate values did not change, and while the electricity rates did increase for those served by GPU, their rates remained within the range charged by other utilities in the state. The four million GPU ratepayers in Pennsylvania and New Jersey were protected from the full impact of the accident by the company's insurance and through extensive contributions, both direct and indirect, from governmental and industry groups to help pay for the cleanup of the damaged reactor and to moderate the costs of purchasing power to replace what would have been produced by TMI. GPU stockholders benefited from this outside help as well because, without it, the company might have gone bankrupt. At the same time, stockholders did sustain considerable costs in the form of lost dividends (estimated at more than $800 million through 1987) and possibly in lower stock prices, though it is still possible that stock prices will fully recover eventually.

These findings have important implications for planners concerned with the possible economic effects of future technologically caused crises. They show that, if there is no physical damage to the surrounding area, the short-term costs will be less than for most natural disasters. They also show that, when the technology involved is complex, expensive, and potentially dangerous, long-term costs can be enormous, often too large for a company and its customers to absorb alone. When this happens, governmental and industry groups are likely to be drawn in, as they were following the TMI crisis. The role that these outside groups play will be critical in determining how the financial impact of such a crisis is distributed among those living in the vicinity of where the accident occurred, customers and owners of the company involved, and society as a whole.

4

Distress and Mediators of Distress Among Persons Living in the Vicinity of Three Mile Island

Overview

This chapter reviews the level and duration of distress experienced by persons living in the vicinity of Three Mile Island over an eighteen-month period following the accident in March 1979. It also reports on the effects of three types of variables expected to mediate the impact of the accident: coping strategies, social support, and respondents' personal characteristics.

There were many studies of distress following the TMI accident, and their findings may, at first, appear contradictory. However, by taking into account when the studies were conducted, the populations studied, the sampling methods used, and the ways in which stress was assessed, a generally consistent pattern emerges. During the crisis, as well as three and nine months later, the general population surveys and most studies of subpopulations showed that a higher percent of persons living close to TMI were upset and reported more stress-related symptoms than did those living farther away. Eighteen months after the crisis, the general population surveys did not show significant differences between persons living close to TMI and those living farther away, though subpopulations that were among those most upset at the time of the crisis continued to show heightened distress near TMI for more than two years after the accident and there is some evidence that it persisted for as long as six years in these subpopulations.

There is evidence that the intensity of the distress near TMI was highest just after the accident and decreased after that. During the crisis period there was evidence of a 10% increase in reports of intense distress (i.e., symptoms that are characteristic of mental patients) among persons living close to the facility. In addition, increased clinical episodes of anxiety and depression were reported by mothers of young children just after the crisis; these episodes gradually decreased in frequency through the following year. The general population surveys conducted in July 1979 and January 1980 (three and nine months after the accident) found that the general level of distress was far below the levels reported by mental patients and in the range of daily stressors. Other investigators' studies of stress after the immediate crisis also found that the intensity of the distress near TMI was far below the levels typically expressed by mental patients.

Most studies of physical symptoms near TMI relied on respondents' self-reports, which can be influenced by many factors in addition to actually experiencing symptoms. For example, it is possible that people living near TMI remembered their symptoms to a greater degree than did people farther away, without actually experiencing more symptoms. If this was the case, then the higher symptom-reporting rates near TMI would reflect greater memory of symptoms rather than greater frequency of symptoms. One of the subpopulation studies used physiological and performance measures of distress, thus avoiding the problems with self-report data. This study also found evidence of heightened distress near TMI, indicating that at least some of the reported distress was physiologically based. At the same time it should be noted that general population studies of physician visits near TMI showed little increase following the accident, indicating that most people living near TMI did not consider their symptoms serious enough to seek medical help.

Social support was found in several studies to be related to lower distress levels. However, there was little evidence that social support buffered the stress experienced as a result of the accident, though one investigator did report evidence for buffering with some stress measures but not with others. Coping strategies, on the other hand, were found by several investigators to be associated with distress following the crisis. Those strategies associated with lower distress included emotion-focused coping, reappraisal, and self-blame, while those associated with higher distress included psychotropic drug use, avoidance behaviors, denial, and problem-oriented coping. In addition, persons who took more protective actions during the crisis and who were active in response to the crisis were more likely to show persistent distress following the crisis. These findings point toward two conclusions. First, people with poor coping strategies (e.g., denial or drug use) were especially likely to experience distress following the crisis. Second, attempts to cope actively (e.g., problem-oriented coping or attending community meetings) with a situation about which the average person could do very little (the technologically and politically complex problems at TMI) increased or maintained, rather than decreased, distress.

There was also evidence that certain demographic and personal characteristics were associated with distress following the crisis. These included being pregnant at the time of the accident, being younger, being married, being female, living in a large household, having a chronic illness, being introspective, being frequently upset, having low self-esteem, and having a prior psychiatric history. Several of these characteristics (e.g., being female, having a chronic illness, having low self-esteem, having a prior psychiatric history) are commonly associated with higher symptom-reporting rates. Their association with distress following the TMI crisis may reflect a general tendency for these groups to report more symptoms as well as sensitivity to the uncertainties and potential for harm associated with the crisis. Other personal characteristics were closely related to the types of people who were advised to evacuate (e.g., mothers of young children, women pregnant at the time of the accident, married persons, persons living in a large household) and therefore reflect who was most vulnerable to harm from the accident.

Background

Research on the Psychological Impacts of Recent Natural and Man-Made Disasters

There is a substantial literature on the psychological impact of disasters, both natural and man-made (Adams and Adams, 1984; Baker and Chapman, 1962; Barton, 1969; Bennet, 1970; Frederick, 1977; Kinston and Rosser, 1974; Lindeman, 1944; Logue and Hansen, 1980; Melick, 1978a, 1978b; Melick, Logue, and Frederick, 1982; Milne, 1977a, 1977b; Perry and Lindell, 1978; Ruskin et al., 1948; Wright and Rossi, 1981). A variety of situations have been studied, ranging from large-scale events, such as floods and earthquakes, to localized but intense events, such as tornados and flash floods. Psychiatric symptomatology has been the most common object of study (e.g., Lifton and Olson, 1976; Rangell, 1976), but researchers have also studied subclinical psychological distress (Kinston and Rosser, 1974; Melick, Logue, and Frederick, 1982; Moore, 1958; Moore and Friedsan, 1959; Perry and Lindell, 1978). The work on the psychological impact of disasters is extensive. For example, Ahearn and Cohen (1983) have compiled a bibliography of about three hundred references pertaining to the relationship between disasters and mental health. Other researchers have investigated physical symptoms (Adams and Adams, 1984; Melick, 1978a, 1978b; Logue et al., 1979), health (Bennet, 1970), effects on family and community (Erikson, 1976a, 1976b; Newman, 1977; Titchener and Kapp, 1976), and individual reactions such as alcohol abuse, violence, and aggression (Adams and Adams, 1984).

Crises most similar to the accident at Three Mile Island are those that are caused by failures of technology. Some examples that have occurred since the TMI crisis include Times Beach, Missouri, where dioxin, a carcinogenic substance, was discovered in the soil; Love Canal, New York, where chemical wastes are suspected of causing health problems; Centralia, Pennsylvania, where a mine fire has created chronic noxious fumes; and Chernoble, in the U.S.S.R., the site of another nuclear accident. It has been suggested that Times Beach, Love Canal, and Centralia have resulted in psychological distress (Holden, 1980; Gibbs, 1983; Reko, 1984), but at the time this book was written there were only a few research studies about the psychological effects

49

of these crises. The most complete studies to date are for Love Canal and Centralia (Levine, 1982; Streufert et al., 1987), which did report evidence of psychological distress.

Unfortunately, there is no consensus on how disasters, natural or man-caused, affect the mental health of individuals (Glenn, 1979; Tierney and Baisden, 1979). Adams and Adams (1984) have reviewed studies that showed little or no negative effect on mental health (e.g., Aquirre, 1980; Hall and Landreth, 1975; Huerta and Horton, 1978; Melick, 1978a, 1978b; Quarantelli and Dynes, 1977; Taylor, 1977) as well as studies suggesting that adjustment problems do occur (e.g., Burke et al., 1982; Logue and Hansen, 1980; Logue, Melick, and Struening, 1981; Newman, 1977; Ollendick and Hoffman, 1982). If symptoms do result, they may be acute and transient in nature (Chamberlin, 1980; Drayer, 1957; Melick et al., 1982; cf. Menninger, 1952; Moore, 1958; Penick et al., 1976; Perry and Lindell, 1978), and some researchers have even suggested that requests for psychiatric services may decline after a disaster (Tierney and Baisden, 1979; Taylor et al., 1979).

It is difficult to know whether the differences in these findings are due to the nature of the disaster, sampling techniques, choice of control samples, measurement of the dependent variable, or time period studied (Drabek, 1970; Penick et al., 1976; White and Haas, 1975). While a number of researchers have proposed models for predicting the impact of disasters (Barton, 1969; Baldwin, 1978; Berren et al., 1980; Chapman, 1962; Glenn, 1979; Green, 1982; Killian, 1954; Quarantelli and Dynes, 1977), these typologies are based on relatively small samples of disasters, most of which are substantially different from the TMI accident on a number of important dimensions. Because of the lack of consistent findings or of a clearly applicable theoretical model, it was not feasible to predict with any degree of certainty the degree or type of distress that would occur following the TMI crisis.

Earlier Research on Mediators of Responses to Stressful Events

Coping Strategies

The way a person interprets and responds to an event may play an important role in how that event affects him or her (Cohen and Lazarus, 1979; Davis, 1963; Katz et al., 1970; Lipowski, 1970; Moos and Tsu, 1977). While there has been a good deal of research on coping, there is little agreement about how individuals react to events and why different events evoke different responses (Silver and Wortman, 1980). However, a serious life crisis may result in distress that persists for a substantial period, and studying the way people respond to events is a critical step in developing an understanding of the residual distress. Lazarus, for example, has emphasized that "to relate stress to maladaptation requires emphasis not so much on stressors as on the cognitive and coping processes mediating the reaction" (Lazarus, 1981:207). He has also argued that the greatest research need at present is for naturalistic research on stress and coping (Lazarus, 1981:207).

Psychological Resources

Psychological resources are "personality characteristics that people draw upon to help them withstand threats posed by events and objects in their environment" (Pearlin and Schooler, 1978:5). There is considerable evidence that the impact of an event on an individual is mediated by such psychological resources (Andrews et al., 1978; Antonovsky, 1979; Kobasa et al., 1981; Chan, 1977; Kaplan, 1970; Kohn, 1976, 1977; Lefcourt, 1981; Smith et al., 1978; Wheaton, 1983). It has been suggested by a number of writers that sense of control (Lefcourt, 1976, 1981) and sense of self-efficacy (Bandura, 1982) are related to how active and persistent people will be in attempting to cope with adverse experiences. Self-esteem (Kaplan, 1970) has also been shown to be related to reduced perceptions of threat (Chan, 1977).

Social Support

The literature on social support is extensive and indicates that, in different circumstances, social support can either have a direct effect on distress or mediate the impact of stressful events (Broadhead et al., 1983; Cohen and Syme, 1985). Thus, for example, a person may be less distressed by a potential evacuation if there are many friends to assist in such an eventuality, or social support may serve to bolster the person's sense of self-worth and increase his or her perceived ability to cope with events (Cobb, 1976).

Studies of Distress after TMI by Other Investigators

A number of studies have been conducted on the psychological stress experienced by the population near TMI following the accident in March 1979. One group of studies was conducted under the auspices of the Presidential Commission on TMI by Dohrenwend, Kasl, and colleagues (Chisholm et al., 1981; Chisholm, Kasl, and Mueller, 1984; Dohrenwend et al., 1979; Dohrenwend et al., 1981; Fabrikant, 1983; Kasl et al., 1981).

Kasl et al. (1981) conducted telephone interviews with 324 nuclear workers at TMI and 298 workers at a comparison plant at Peach Bottom. TMI workers thought they had greater radiation exposure than did Peach Bottom workers and that their health had been endangered as a consequence. They also reported more uncertainty and conflict at the time of the accident, and they were more likely to evacuate after the accident. Forty percent of the TMI workers wished to leave but could not because of work obligations. The workers at TMI reported much lower job satisfaction and greater uncertainty about their job future, but coping responses (such as seeing a doctor, taking drugs, and increased alcohol consumption) were infrequent. At the time of the accident, workers experienced more periods of anger, extreme worry, and upset and more psychophysiological symptoms, with some of these feelings and symptoms persisting for at least six months after the incident. Psychological demoralization was greatest among TMI nonsupervisory workers, and the effect was not related to how far from the plant the worker lived. Presence of children at home was related to greater impact of the incident, but mainly among TMI supervisors.

Studies of social support among TMI workers (Chisholm, Kasl, and Mueller, 1984) indicate that both supervisor and coworker support were related to less distress and more job satisfaction. However, social support did not necessarily reduce negative reactions or increase positive results. Specifically, it consistently affected the impact of events at low levels of distress but not at high levels.

Dohrenwend and colleagues (Dohrenwend et al., 1981) also studied teenagers in the seventh, ninth, and eleventh grades from a school district within a twenty-mile radius of TMI, mothers of preschool children from the same area, a similarly drawn control sample from Wilkes-Barre (which is about a hundred miles away), and a sample of male and female heads of households from the same area (within twenty miles). They found that demoralization among heads of households near TMI was sharply elevated in April, immediately after the accident, but that scores fell sharply in a second sample in May and showed a small additional drop in a third sample in July. A comparison of demoralization scores of mothers with small children in the vicinity of TMI and in Wilkes-Barre showed no significant difference in July. They estimated that, among persons living near TMI immediately after the accident, there was a 10% increase over what would normally be expected in the number of persons with demoralization scores higher than the mean scale scores of persons receiving treatment in a mental health clinic. They also found that the "distrust of authority" scale was markedly elevated immediately following the accident and that it dropped slightly in May, with a slight additional drop in July. However, they also found in July that distrust was higher than in a control group in Wilkes-Barre, and, in addition, it was also higher than would be expected when compared to the results of a national survey using similar questions.

The Dohrenwend group also found that perceived threat to physical health from the TMI accident was higher in the general population near TMI immediately after the accident but was much lower by July. Household heads living within five miles of TMI were more uncertain about whether their health had been affected by the accident than those living farther away, and mothers of preschool children in the area also felt more uncertain about whether their physical health had been affected than were mothers of preschool children in

52

Wilkes-Barre. The studies of seventh-, ninth-, and eleventh-grade students living near TMI showed that children who reported higher distress during the accident lived closer to TMI, had a preschool sibling, and evacuated the area during the crisis. In addition, those with a preschool sibling and who left the area during the crisis also reported elevated frequencies of low intensity somatic symptoms and greater persistence of distress a month after the accident.

Goldsteen and Schorr (1982) reported persistent concerns about the situation at Three Mile Island through March 1980 among a sample of residents of Newberry Township and Goldsboro, Pennsylvania, which are within five miles of Three Mile Island. They also reported SCL90 scores for their samples in October 1979 and in March 1980 that were above those obtained by the Presidential Commission study for the same area shortly after the crisis. However, it should be noted that their study did not include a matched control group.

Bromet and her colleagues conducted a series of studies (Bromet and Dunn, 1981; Bromet et al., 1982a; Bromet et al., 1982b; Bromet, Hough, and Connell, 1984; Bromet and Schulberg, 1984; Dew, Bromet, and Schulberg, 1987a, 1987b; Fienberg, Bromet, and Follmann, 1985; Parkinson and Bromet, 1983; Solomon, 1985; Solomon and Bromet, 1982) concerned with the impact of the crisis on mothers of preschool children, workers at TMI, and community Mental Health System clients.

Of the three groups, mothers of young children were the ones most affected by the accident. Bromet and her colleagues studied 385 mothers near TMI and compared their responses to those of a comparable group living near a nuclear power plant in Beaver County, Pennsylvania, approximately two hundred miles away. The TMI mothers, when compared to the control group, had an excess risk of reporting clinical episodes of anxiety and depression, as well as more symptoms of anxiety and depression at subclinical levels, during the year after the accident. The clinical episodes of anxiety and depression occurred primarily in the period shortly after the accident, which is consistent with the Dohrenwend group's findings discussed above. Subclinical levels of anxiety and depression were also reported to be elevated among the TMI mothers in interviews at 9, 12, 32, and 42 months after the crisis as well as at the time of restarting the undamaged reactor at TMI (six years after the crisis). However, it should be noted that only at 9 and 12 months was distress significantly greater than in control groups. At 30 and 42 months the mothers in the control group showed a marked rise in distress, while the TMI mothers remained at the same levels as in earlier assessments. The investigators concluded, based on interviewers' reports and other analyses, that the elevated unemployment rate in the control group (which reached 46% at 42 months) was responsible and that it therefore no longer qualified as a low-stress control group against which to compare the TMI mothers. Therefore, their conclusion that the TMI mothers remained distressed for up to 42 months after the accident was based on the observation that their stress levels did not markedly change after the 12-month interviews.

The stress levels among mothers living near TMI were studied again by the Bromet group when the undamaged reactor at TMI was restarted six years after the original accident (Dew et al., 1987b). During those six years the restart had been the center of public controversy, which included allegations that the undamaged reactor was unsafe and that the mental health of people living in the area would be jeopardized by the restart. In this study the investigators compared SCL90 scores obtained in interviews in the fall of 1981 and 1982 with those obtained from the same women just after restart in the fall of 1985. They found that their scores were significantly higher in 1985 than in the 1981–82 interviews and reasoned that restart was the likely cause, although this cannot be determined with certainty without a control group. (A similar study by Prince-Embury and Rooney [1987] also reported that SCL90 scores of persons near TMI were elevated following restart when compared to normative data, but, as with the Dew et al. study, there was no matched control group.)

Factors related to distress in the various studies conducted by the Bromet group were having a prior psychiatric history, living within five miles of the plant, having less adequate social support, and being pregnant at the time of the accident. However, the data did not support the hypothesis that social support buffered the impact of TMI-related distress (Solomon, 1985). Bromet, Hough, and Connell (1984) also studied 150 children from the TMI area and 99 children from a comparison site three and a half years after the accident, but they found no significant differences between the two groups in terms of social competence, behavior problems, fearfulness, or self-esteem.

Bromet et al.'s study of workers at the Three Mile Island facility showed that, when compared with a control group, TMI workers had more mental health problems, but there were preexisting differences between the groups. Interestingly, TMI workers felt more rewarded by their jobs than workers at the comparison site. In both the TMI and control workers, social support was related to fewer symptoms and greater feelings of reward from the job, but it did not appear to moderate the effects of occupational stress.

Bromet also found that mental health clients living near TMI who perceived the accident as dangerous and felt living near a nuclear facility was unsafe had consistently higher anxiety scores. However, she found that symptom levels were not higher than in the control group of clients from Beaver County, Pennsylvania, after statistically controlling for education, age, and residential mobility.

Baum and colleagues (Baum, Fleming, and Singer, 1983b; Baum, Gatchel, and Schaeffer, 1983; Collins, Baum, and Singer, 1983; Davidson and Baum, 1986; Fleming et al., 1982; Gatchel, Schaeffer, and Baum, 1985; Schaeffer and Baum, 1984) conducted a study before, during, and after the venting of

54

krypton gas from the damaged reactor at Three Mile Island. Data was collected at four times, beginning fifteen months and ending seventeen months after the accident. Data was collected from fifty-four residents within five miles of TMI and several samples of control subjects away from the plant. The number of TMI subjects in different publications varies from thirty-eight to fifty because not all measures were obtained from all respondents. They found that subjects living within five miles of TMI exhibited greater stress on psychological, physical, and behavioral measures than control subjects at all four data collections, including seventeen months after the crisis. At fifty-eight months postcrisis they conducted another study (Davidson and Baum, 1986), in which a sample of fifty-two persons living near TMI was compared to a control group in Frederick, Maryland. The TMI sample consisted of persons who participated in the earlier study plus additional subjects recruited in the same manner as the original group. Davidson and Baum reported significant differences between the TMI and control groups on the same measures as in the earlier studies. They also investigated the prevalence of symptoms similar to those characteristic of the posttraumatic stress syndrome and found that the TMI group reported more such symptoms than did the control group.

Some of the most important contributions of the Baum group's studies to understanding the impact of the crisis were their physiological findings. Other investigators had to rely on respondents' self-reports of physical symptoms (which are vulnerable to response biases), but the Baum group measured catecholamine levels in the TMI and control samples. Catecholamines, which have been shown to vary with stress, were elevated in the TMI samples as long as fifty-eight months after the crisis. They also assessed performance on cognitive tasks, which have also been shown to be sensitive to stress, and found the TMI sample performed less well than the controls. Both of these stress measures, as well as scores on the SCL90 scale, indicated that the intensity of the long-term stress near TMI was subclinical, that is, below the levels characteristic of mental patients and within the range of everyday stressors.

They also studied the mediating influences of different coping styles and found that emotion-focused coping and self-blame were associated with fewer reported symptoms and that greater use of problem-oriented coping was associated with greater symptom reporting (Baum et al., 1983b; Collins et al., 1983). In addition, they found that TMI residents who reported use of denial reported more symptoms, and use of reappraisal was associated with less symptom reporting. Collins, Baum, and Singer (1983) suggest that, because the stress associated with the TMI incident was chronic and the sources of stress were not easily changed, reappraisal-based emotion management was most effective. These features of the stressor also can explain why denial or problem-oriented modes of coping were less useful in reducing stress.

The Baum et al. group investigated the role of social support and found that having more social support was associated with fewer stress-related problems, but that the intervening effects of support were inconsistent across psychological, behavioral, and biochemical measures. Fleming et al. (1982) interpret these results as suggesting that support serves to facilitate coping rather than to protect people from stress. That is, under conditions of high support, subjects were still "stressed" and aroused, but they may have been better able to cope and to minimize the aversiveness of stress.

Mileti, Hartsough, Madson, and Hufnagel (1984) analyzed data that were not as subject to reporting biases as self-reported stress symptoms, such as cardiovascular deaths, criminal arrests, psychiatric admissions, suicides, automobile accidents, and alcohol sales for six months prior to the accident and six months after the accident among residents within five miles of TMI, from five to ten miles of TMI, and a matched control group from Wilkes-Barre, Pennsylvania. Rates of crime, psychiatric admissions, and suicide did not show significant differences between the TMI and control groups. However, in the first few days after the accident, traffic accidents were more frequent in the five-to-ten-mile ring and there was an increase in alcoholic beverage sales. Walsh and Warland (1983) investigated the characteristics of persons who coped by contributing time or money to anti-TMI citizen protest groups and found that, among those who were anti-TMI, activists tended to be of a higher socioeconomic status than the nonactivists, to have been more politically active before the crisis, and to have liberal political views.

There have been numerous other, nonscientific descriptions of the population's reaction to the incident (e.g., Mills, 1980). These reports also suggest that the incident resulted in elevated levels of distress that persisted, for some people, for as long as a year.

Distress Following the Accident as Shown in the General Population Surveys

Limitations of Self-Reported Data

The distress data obtained in the NRC, PDH, and RWJ telephone surveys (see Table 1.2 for explanations of surveys) included ratings of how upset respondents were about the situation at TMI as well as reports of symptoms for different two-week periods. While respondents' statements of how upset they were about TMI can be taken at face value, care should be taken in interpreting the symptom-reporting data. Many factors in addition to a respondent's actually experiencing symptoms can affect symptom reporting. First, the questions require retrospection and hence are subject to selective recall and forgetting.

Second, it is well established that, under certain circumstances, a general willingness to acknowledge feelings and symptoms can affect responses to questions about symptoms. Third, people sometimes remember and report events in a manner that is consistent with their attitudes (Festinger, 1957) or behavior (Bem, 1972). This could lead people with negative attitudes or those who had evacuated during the crisis to remember and report more stress-related symptoms than people with positive attitudes or those who did not evacuate. In addition, it is possible that some respondents consciously distorted their answers in hopes of affecting public policy toward TMI. Finally, and perhaps most important, reporting of physical symptoms is often related to psychological distress. Thus, being upset emotionally may lead individuals to focus attention on or magnify their physical symptoms.

Despite these limitations, reported symptoms are often an excellent indication of the strain a person is experiencing in a given situation. Even if respondents magnified their symptoms because they were upset, or consciously overreported symptoms in hopes of influencing TMI policy, the heightened symptom reporting indicates serious concerns on the part of those reporting the symptoms. For our purposes, then, it is not necessary to assume that all of the reported symptoms actually occurred in order to say that higher rates of symptom reporting close to TMI as compared to farther away indicate that people living near the reactor were upset about TMI.

Concern about TMI

Results for the question of how upset respondents were about TMI are reported in Figures 4.1 and 4.2. Data are presented in two ways. In Figure 4.1, we present the percent of persons within each distance group who reported being extremely or quite upset about the situation at Three Mile Island. This figure shows a sharp overall drop from April 1979 to January 1980, though symptoms are higher close to TMI at both times. It is interesting to note that, even in the farthest group, more than 20% were extremely or very upset about TMI in April, and 7% still felt that way in July. This indicates that distress was experienced to some degree even in the farthest group.

These same data were analyzed using multiple regression analyses to control for a number of demographic variables (age, sex, education, income, and marital status) that differed somewhat at different distances, hence possibly influencing the results. These adjusted results are shown in Figure 4.2 in terms of variations from the farthest group (beyond forty miles). Figure 4.2 shows a marked distance effect in April 1979 (immediately following TMI), as well as in January 1980, though the percentages within fifteen miles for January are half what they were in April. In April the probability of being extremely or very upset is significantly higher for the 0–5, 6–10, and 11–15 mile groups

FIGURE 4.1

PERCENT OF RESPONDENTS AT DIFFERENT DISTANCES FROM TMI WHO WERE EXTREMELY UPSET OR QUITE UPSET DURING THE TMI CRISIS, IN JANUARY, 1980 AND, IN OCTOBER, 1980*

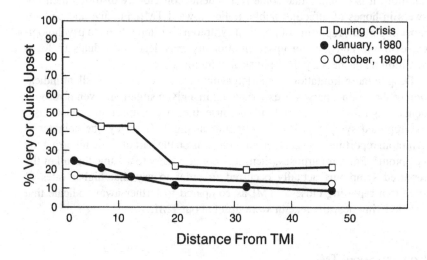

*The six plotted points represent the percent of respondents within the following areas as measured from Three Mile Island: 0-5 miles, 6-10 miles, 11-15 miles, 16-25 miles, 26-40 miles and 40-55 miles.

than for the farthest group. In January they are significantly different for only the 0–5 and 6–10 mile groups. In the survey conducted in October 1980 (Houts and Goldhaber, 1981), the upset ratings of the distance groups were no longer significantly different.

Stress-related Symptoms

Data on stress-related symptoms were obtained in all of the telephone surveys. The symptoms, which included stomach trouble, headaches, diarrhea, constipation, frequent urination, rash, abdominal pain, loss of appetite, overeating, trouble sleeping, sweating spells, feeling trembly and shaky, trouble thinking clearly, irritability, and extreme anger, were drawn from the general literature on psychophysiological symptoms of stress (e.g., Selye, 1956), as well as from interviews with more than two hundred patients at the Hershey

FIGURE 4.2

CORRECTED* PERCENT OF RESPONDENTS AT DIFFERENT DISTANCES FROM TMI WHO WERE EXTREMELY OR QUITE UPSET DURING THE TMI CRISIS, IN JANUARY, 1980, AND IN OCTOBER, 1980

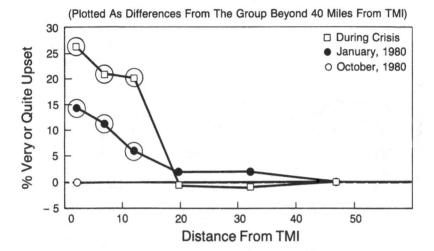

(Plotted As Differences From The Group Beyond 40 Miles From TMI)

*Age, sex, education, income and marital status controlled. A circled point indicates that the percent of persons who were extremely or very upset about TMI at that distance was significantly greater than the percent of persons beyond 40 miles who were extremely or very upset.

The six plotted points represent the percent of respondents within the following areas as measured from Three Mile Island: 0-5 miles, 6-10 miles, 11-15 miles, 16-25 miles, 26-40 miles and 40-55 miles.

Medical Center, Hershey, and in the practice of Dr. Joseph Leaser, Middletown, immediately following the crisis. Respondents were asked if they had experienced each of these symptoms anytime in a two-week period.

Reported frequency of symptoms experienced is summarized in Figure 4.3. (Analyses utilizing subgroups of these symptoms can be found in reports to the Pennsylvania Department of Health [Houts et al., 1980], where the same patterns reported here can be seen.) The data in Figure 4.3 indicate large overall differences for April, July, and January in the percent of persons who reported at least one of these symptoms. The highest rates are reported in January, followed by July and April. The fact that July rates are higher than those for April is probably due to memory decay, since both sets of data were

FIGURE 4.3

PERCENT OF RESPONDENTS AT DIFFERENT DISTANCES FROM TMI WHO REPORT ONE OR MORE STRESS-RELATED SYMPTOMS DURING THE TMI CRISIS, IN JULY, 1979, IN JANUARY, 1980, AND IN OCTOBER, 1980*

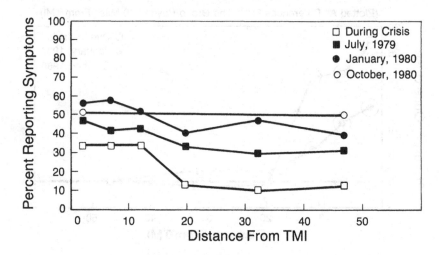

*The six plotted points represent the percent of respondents within the following areas as measured from Three Mile Island: 0-5 miles, 6-10 miles, 11-15 miles, 16-25 miles, 26-40 miles and 40-55 miles.

collected in July. When reporting April symptoms, respondents had to remember back three months, but in answering about July, they had to remember only for the two weeks just prior to the interview. The fact that rates are higher in January than in July is probably due to seasonal variations in symptom reporting, rather than to memory, since both involved memory for only the preceding two weeks. The Health Interview Survey, a nationwide survey conducted by the National Center for Health Statistics, reports that acute symptoms are, on the average, 1.4 times higher in January than in July (National Center for Health Statistics, 1979). This is approximately the same as the differences found here.

These findings indicate that those living closer to TMI reported experiencing more stress-related symptoms than those living farther away. Figure 4.4 shows this more convincingly because demographic variables (i.e., age, sex, education, income, and marital status) have been controlled and probabilities are plotted as deviations from the control group (beyond forty miles). The results,

FIGURE 4.4

CORRECTED* PERCENT OF RESPONDENTS AT DIFFERENT DISTANCES FROM TMI REPORTING ONE OR MORE STRESS-RELATED SYMPTOMS DURING THE TMI CRISIS, IN JULY, 1979, IN JANUARY, 1980 AND IN OCTOBER, 1980

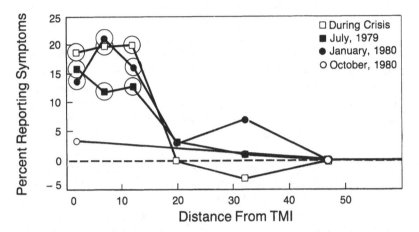

*Age, sex, education, income and marital status controlled. A circled point indicates that the percent of persons reporting symptoms at that distance was significantly greater than the percent of persons beyond 40 miles reporting symptoms.

The six plotted points represent the percent of respondents within the following areas as measured from Three Mile Island: 0-5 miles, 6-10 miles, 11-15 miles, 16-25 miles, 26-40 miles and 40-55 miles.

in Figure 4.4, indicate not only raised response frequencies close to TMI but also a sharp drop between fifteen and twenty-five miles, the same pattern seen with the upset data. Statistical significance, as shown in Figure 4.4, also drops after fifteen miles. The increase in symptom reporting is approximately 15% over what was reported forty to fifty-five miles from TMI, all three times. By October 1980, however, the differences between those living close to the reactor and far away were negligible and not statistically significant.

Psychological Symptoms: The Langner Index

The Langner Index of psychological distress was also included in the January RWJ survey, which extended out to fifty-five miles from TMI. This scale consists of questions about relatively severe symptoms of depression and anxiety and some psychosomatic symptoms that are characteristic of persons receiving treatment for mental illness. Data were analyzed in the same manner as for the other distress indices. However, there were no statistically significant differences between groups close to TMI and the farthest comparison group in January 1980, nine months after the crisis.

The fact that the Langner scale shows a different pattern from the other distress measures suggests that it is measuring a different degree of distress or a different type of distress than are the questions about upset or the stress-related symptoms. A detailed analysis of each of the items in the Langner Index and the stress-related symptoms supports this interpretation. The Langner scale includes questions about depression as well as psychosomatic symptoms not included in the stress-related symptom lists. In addition, even where the symptoms overlap, the scoring has a higher threshold for the Langner scale. Most questions in the Langner Index allow for three responses: never, sometimes, or often. The usual Langner scoring of these items is to count only "often" as a positive response. When this is done, none of the items in the Langner Index shows an increase close to TMI. However, when the response "sometimes" is also included, a procedure that makes the scoring more comparable to the PDH and NRC studies, a distance effect is seen, but only for those items that overlap with those in the stress-related symptom lists, which also showed a distance effect. This indicates that differences in both type and severity of symptoms assessed contribute to the different findings with these two distress measures. It also indicates that, since the Langner Index was designed to assess stress at the level characteristic of mental patients, the long-term distress resulting from living near the TMI facility was below the level characteristic of mental patients. This conclusion is the same as that of both Bromet et al. (1982a) and of Schaeffer and Baum (1984), who reported elevated symptom reporting on the SCL90 scale among persons living near TMI, but also that these levels were below those of mental patients.

Physician Visits

One potential indicator of impact is physician visits. People might use more physician services because of new symptoms after a traumatic event. Stress may sensitize people to symptoms that, in turn, may cause them to see a physician, or people may visit a physician as a way of coping with distress (Mechanic, 1972). Often, physical complaints may be presented to justify such visits, but the underlying reason may be psychological (Balint, 1957; Clyne, 1961).

To investigate how the TMI accident affected the use of health care services, we analyzed Blue Cross/Blue Shield records of claims by 556 primary care physicians in the vicinity of TMI, utilization rates prior to and subsequent to the TMI crisis in a family practice located near Three Mile Island, and opinions of primary care physicians practicing within twenty-five miles of TMI about how the crisis had affected their practices (Houts et al., 1984a). The Blue Cross/Blue Shield data for primary care physicians practicing in the five counties surrounding TMI showed a small, nonstatistically significant trend for physicians within fifteen miles of TMI to increase claims following the accident when compared to those practicing beyond fifteen miles. The family practice data included practice utilization rates of 498 patients before and after the crisis. Again there was a small, nonstatistically significant trend for patients who reported being upset during the crisis to increase their utilization after the crisis when compared to patients who were not upset during the crisis. The survey of 223 primary care physicians within twenty-five miles of TMI indicated that in approximately five patient visits per physician the patients believed their problems were due to stress from TMI. However, in the physicians' judgments, four out of five of these medical problems were due to other causes. The family practice study did yield one statistically significant finding of interest. It was found that persons who were upset during the crisis tended to be high users of medical services both before and after the crisis. This suggests that the type of person who makes frequent use of physician services is also the type of person who responds more emotionally to stressful situations, which is consistent with the established association between psychological distress and use of medical and psychiatric services (Mechanic, 1978).

Summary and Integration of Findings on Distress Following the TMI Crisis

The general population survey findings for upset and for stress-related symptoms are strikingly similar. Both sets of data show raised levels of distress out to fifteen miles, and both indicate essentially the same pattern for the April, July, and January surveys. While respondents' statements about how upset they were about TMI can be taken at face value, we cannot be certain about the meaning of the symptom reports, for the reasons mentioned earlier. Nonetheless, some general conclusions are possible from the symptom-reporting findings. As discussed in the introduction to this chapter, the Baum et al. group reported both physiological evidence of increased stress near TMI and deficits in performance on cognitive tasks in their TMI sample. Those data are not subject to the possible response biases that can affect self-reported physical symptoms. While the Baum et al. sample was small, their control groups were closely matched to their TMI group, and so it can be concluded that at least

63

some of the elevated symptom reporting near TMI was physiologically based. On the other hand, the data from physician surveys suggests that some people attributed symptoms to TMI that had other causes, suggesting that some of the symptom-reporting may have been because people who were upset about TMI noticed and reported symptoms more frequently. It is likely that both factors played a role in the general population survey findings. At the same time, even if some of the symptom reports were not physiologically based, they still have important implications for understanding the impact of the crisis. If the higher reporting rates near TMI are due partially to strong negative attitudes toward the plant, causing people to notice and remember symptoms, this too indicates higher distress near the island.

Most of the findings by different investigators can be shown to follow a consistent pattern by taking into account the timing of the investigations, the populations they studied, and how they assessed stress. In general there was intense stress during the crisis, dropping to low intensity distress after the crisis and continuing to slowly drop so that persistent low level stress was evident in the general population surveys for nine months after the accident and in certain subpopulations, which were most distressed at the time of the crisis, for several years after the accident.

We will first consider the studies reported by the Presidential Commission and conducted by the Dohrenwend et al. group. These studies showed evidence of severe stress just after the accident, but in July 1979, three months later, they found no differences in the stress levels of mothers of young children near TMI and a comparable group living in Wilkes-Barre, one hundred miles away. On the other hand, they did find heightened distrust of authority near TMI in their July study. There is agreement between the general population surveys reported here and the Presidential Commission findings for the period just after the crisis. Both studies found elevated distress near TMI, and the Presidential Commission studies showed that it was severe enough to increase the reporting of symptoms characteristic of mental patients. However, there is disagreement between the general population findings of higher rates of stress-related symptoms near TMI in July 1979 and the Presidential Commission findings of no differences in stress between mothers of young children near TMI and one hundred miles away in July 1979. In addition, there is an apparent disagreement between the Presidential Commission findings and those of Bromet et al. showing heightened distress among mothers of young children near TMI compared to a matched control group several hundred miles away nine months after the crisis and the Baum et al. studies showing heightened distress near TMI fifteen months after the crisis.

The study closest to the Presidential Commission sample of mothers is the Bromet investigation of mothers of young children conducted approximately nine months after the crisis. The sampling methods were essentially the same in the two studies, and the instruments used to assess stress—namely, the demoralization scale and the SCL90 scale—are similar in content and format and are highly correlated. There are several possible reasons for the differences in their findings, including sampling error, differences in how the questionnaires were administered (Bromet's were face-to-face interviews, while the Presidential Commission's were not), and differences in time period for which symptoms were reported (the two to three months since the TMI accident in the Presidential Commission study, as compared to one month in the Bromet et al. study). In addition, there is the possibility, suggested by Dohrenwend (personal communication), that the stress levels for mothers of young children may have dropped shortly after the accident and then increased nine months later. This is not consistent with the NRC survey findings of 15% higher symptom-reporting rates near TMI than forty to fifty-five miles away at approximately the same time that the Presidential Commission study was surveying mothers of young children. However, because the NRC and Presidential Commission studies used different questions and sampling methods, it is not possible to definitively resolve this issue.

It should be noted that the report the Dohrenwend group submitted to the Presidential Commission included a discussion of both the Demoralization scale findings and the findings of persistent distrust of authority near TMI and recognized the possibility of some long-term psychological effects of the accident on people living near the reactor. Unfortunately, the summary report of the Presidential Commission, which was *not* written by the investigators, included only findings from the Demoralization scale in its discussion of long-term psychological effects and ignored findings about persistent distrust in authority. As a result, its conclusion that "there was immediate, *short lived* [emphasis ours] mental distress produced by the accident among certain groups of the general population" was oversimplified and incomplete.

Mileti et al. (1984) studied cardiovascular deaths, criminal arrests, psychiatric admissions, suicides, automobile accidents, and alcoholic beverage sales near TMI following the accident. They found little evidence of the accident's impact other than increased traffic accidents and alcoholic beverage sales in the first few days after the accident. How do these findings relate to the findings on long-term distress in the general population surveys as well as those by the Bromet and Baum groups? First, it should be noted that Mileti et al.'s findings on psychiatric admissions are consistent with the Langner scale findings in the general population surveys and with Bromet's findings for patients in mental health clinics, both of which failed to show an impact of the crisis. Second, most of the distress measures used by Mileti et al. were "high threshold" measures. That is, the distress had to be in the severe range to affect those indices. Their finding that these measures did not show any long-term impact of the accident is consistent with the conclusions of Bromet et al. and Baum et al., as well as ours based on the general population surveys, that the long-term stress effects of the TMI crisis were of low intensity.

Finally, there is the apparent disagreement between the general population survey findings and the findings of both Bromet et al. and Baum et al. on how long stress persisted after the TMI accident. In October 1980, eighteen months after the crisis, the general population surveys failed to show significant differences in either symptom reporting or being upset about TMI between people living within five miles of TMI and those living forty to fifty-five miles away. However, the Baum group reported significant differences, seventeen months after the accident, between people near and far from TMI on a number of stress measures. Subsequently both the Baum and Bromet groups reported evidence of heightened, though low intensity, stress among persons near TMI six years after the crisis. The differences between these findings can be reconciled, we believe, by considering the characteristics of the populations studied. The general population studies were samples of all households with phones within five miles of TMI and forty to fifty-five miles away. They therefore include representatives of the many different types of persons and families that are found in the general population. The Bromet et al. and Baum et al. samples are more specialized. The Bromet et al. sample was restricted to mothers of children who were preschool age at the time of the accident. As will be shown later in this chapter, the general population studies showed that these were among the most upset during the crisis. The Baum et al. sample was skewed toward younger female respondents with children, who, as will be shown later in this chapter, were associated with greater reports of distress following the accident. Baum et al.'s (1985) report of findings from physician record data is another indication that their sample was more affected by the crisis than the population as a whole. They compared medical records of eighteen persons near TMI who had visited their physicians at least once in the year prior to and after the accident to a comparable control group of sixteen persons in Frederick, Maryland. They found that, after the accident, there was a significant increase in the problems recorded and prescriptions issued for the TMI group, which did not occur in the control group. In contrast, our analyses of all Blue Cross/Blue Shield claims near TMI, as well as our analyses of the records of 498 persons in a family practice near TMI, showed little impact of the crisis, suggesting that people in the Baum et al. sample reacted more strongly to the crisis than did the population as a whole. Our conclusion is that both the Bromet et al. and the Baum et al. samples differed from the general population in that they contained more persons likely to be especially upset by the accident and the dangers it posed.

These differences in the populations studied by the different investigatory groups suggest the following explanation for the differences in their findings. Distress levels in the general population were initially high, but gradually diminished, as shown in the general population surveys three, nine, and eighteen months after the crisis, to the point where only certain subgroups were experiencing continuing distress. The numbers of persons in those subgroups were relatively small, so that they did not have a major impact on the general population means. One way to check on this interpretation would be to examine, in the October 1980 general population survey, only mothers of young children who were living in the same residences as at the time of the TMI accident. We did this and found a trend in the predicted direction, but it was not statistically significant. However, there were only ten respondents in the control group forty to fifty-five miles from TMI who met these selection criteria, and therefore the likelihood of a type II error (a true difference not being statistically significant) was large.

Characteristics of Persons Experiencing Distress

Several features of the TMI accident and its aftermath suggest that people's psychological characteristics influenced the levels of stress they experienced following the accident. First, in contrast to most natural disasters, there was no detectable physical damage or illness as a direct result of the incident (Upton, 1981). Therefore, any impact of the incident was mediated by psychological reactions or behavioral responses, such as evacuation. Second, the nature, severity, and duration of the threat were ambiguous. Third, whereas in most disaster situations the danger ceases at a relatively specific time, many residents in the vicinity of Three Mile Island perceived a continuing threat associated with the reactor, even after the crisis was declared over. These factors make the TMI accident a unique opportunity for studying how personal characteristics and differences in coping responses affect the strength and persistence of psychological distress.

Characteristics Studied

In the general population studies, four sets of characteristics were examined in relation to the way in which people reacted to the TMI incident. First, distance from the reactor, an objective measure of relative danger associated with a nuclear mishap, was included in the multivariate analyses to control for objective danger when examining relationships between other respondent characteristics and levels of distress. Two other variables that are related to distance and might be expected to affect distress were also included: 1) whether respondents could see the reactor from where they lived, and 2) whether they could see the reactor from their place of work. Those who could see the reactor might be continually reminded of the danger, which, in turn, could increase their distress levels.

A second set of variables examined was sociodemographic characteristics, including age, sex, education, income, and whether the respondent was married. A third set of variables was indicators of vulnerability and included having chronic health problems as well as variables that indicated similarity to persons advised to evacuate: whether respondents lived in a house where someone was pregnant or where there was a preschool child, and the size of the household.

Other variables studied were indicators of the respondents' emotional responsiveness: the frequency with which the respondents said they get upset and their propensity to express their feelings openly. It was expected that more expressive people would say they were more upset and report more psychophysiological symptoms because of their general tendency to express their concerns and worries.

Analyses

Multiple regression analysis was used to test which respondent characteristics were associated with each of the distress measures. The measures of distress used for these analyses were: 1) reported upset, and 2) symptoms reported during the TMI crisis. The results, shown in Table 4.1, indicate that a number of the characteristics studied were associated with response to the crisis. As expected, distance from the reactor was a predictor of emotional reactions. In the NRC study, which extended to fifty-five miles from TMI, those living closer to the reactor were more upset by the incident and reported more psychophysiological symptoms. In the PDH study, being able to see the reactor while at work was also associated with a tendency to report symptoms.

Several characteristics that indicated similarity to persons advised to evacuate during the crisis were found to be related to how upset respondents were about the situation at Three Mile Island. Married people were more upset than those who were not married. Respondents in large households were more upset than those in small households, and younger respondents were more upset than older respondents. However, neither marital status nor age was significantly related to how many symptoms people reported. Women were more upset about the incident and reported more psychophysiological symptoms than did men. The number of chronic health conditions reported was related to the number of symptoms reported, which is probably a reflection of respondents' general health. At the same time it was also related to how upset the respondent was about TMI, which may reflect vulnerability to stress among persons with chronic illnesses. Finally, as expected, respondents' reports of how frequently they get upset were related significantly to both degree of upset and number of symptoms reported, and the reported tendency to express feelings openly was associated with being more upset.

The association between seeing the reactor when at work and symptom reporting is likely the result of being continually reminded of the danger. This

Table 4.1

Characteristics of Persons Reporting Stress during the TMI Crisis (Standardized Regression Coefficients)

Independent variables	NRC Study 0–55 miles from TMI		PDH Study 0–5 miles from TMI	
	Upset	Symptoms	Upset	Symptoms
Proximity				
Miles from reactor	−.24**	−.15**	X	X
Can see reactor at home	X	X	−.02	−.03
Can see reactor at work	X	X	.05	.12**
Sociodemographic				
Age	−.19**	−.06*	−.15**	−.06
Sex (female = high score)	.18**	.11**	.21**	.10**
Education	.08**	−.01	.09*	.005
Income	−.07**	−.07*	−.03	.002
Married***	.09**	−.02	.08*	−.01
Vulnerability/Support				
Household size	.01	−.03	.12**	.006
Someone in household pregnant	−.01	−.02	.01	.05
Someone under 6 yrs in house	.01	.03	.003	−.03
Number of chronic health conditions	X	X	.15**	.33**
Emotionality				
Gets upset frequently	X	X	.11**	.10**
Expresses feelings openly	X	X	.09**	−.03
R2	.15	.05	.18	.15
F value	25.81**	7.75**	10.17**	8.61**
N	1486	1486	692	692

*p < .05.
**p < .01.
***Compared to all nonmarried respondents.
X = variable not included in study.

could have increased the rate at which stress-related symptoms were experienced or, as explained earlier, increased awareness of and memory for symptoms. In either case it indicates that Three Mile Island had a continuing effect on persons living in its vicinity. The relationship between the number of symptoms reported and certain characteristics, such as chronic illnesses or sex of the respondent, is, of course, not necessarily a function of the incident and might be expected in any population, though it may also indicate vulnerability to the stresses of the crisis as well. The association between being upset about TMI and the types of persons advised to evacuate is not surprising since such persons were publicly identified as the group most at risk. Finally, it should be noted that tests for the interaction between distance and significant predictors yielded no significant results.

Mediators of Stress Reactions

Coping Strategies

A number of specific coping responses were evaluated in the PDH July survey within five miles of TMI. Respondents were questioned regarding specific protective actions they took for their families (e.g., kept family indoors, sent family out of the area, had tests for radiation, changed diet, changed use of cow's milk). Questions were also asked concerning changes in drug, alcohol, and tobacco use, evacuation behavior (and reasons for evacuating or not evacuating), and political involvement. Additionally, respondents were questioned regarding a list of specific coping strategies. Since evacuation was dealt with in depth in Chapter 1, we do not reexamine this particular coping strategy.

Our measurement of coping responses began after the occurrence of the stressful event and probably after all participants had experienced some effects of the event. Therefore, it was not possible to study the roles of coping and social support in buffering stress during the crisis period. Instead, we examined the extent to which certain types of social resources were related to the persistence of distress and perceived symptoms after the crisis.

Shortly after the accident there was considerable publicity concerning the possibility of milk contamination due to possible radioactive fallout from Three Mile Island. Despite this publicity, only 8% of the respondents within five miles of TMI reported any change in their use of cow's milk. Of these, 27% reported stopping the use of milk altogether, 15% changed to powdered milk, and 18% changed to milk from a distant source; the remaining 40% carried out some other change. The only other protective measure taken by a significant proportion of the population was to remain indoors as much as possible during the period of immediate crisis. Few respondents reported either having radiation tests or changing their diets. (See Table 4.2.)

Respondents within five miles of TMI were also asked about changes in their use of alcohol, tobacco, sleeping pills, and tranquilizers during the period just after the accident. Generally, increases were seen for all four substances. (See Table 4.3.) Fourteen percent of respondents who regularly drank alcoholic

Table 4.2
Other Protective Actions

Respondents reporting change in use of cow's milk	8%
Respondents reporting doing the following things to protect family:	
Kept members inside	56%
Had radiation tests	6%
Changed diet	3%

Table 4.3
Changes in Use of Alcohol, Tobacco, Sleeping Pills,
and Tranquilizers during Crisis Period

Proportion of drinkers reporting increased alcohol use	14%
Proportion of smokers reporting increased smoking	32%
Ratio of people using sleeping pills during the crisis to those using sleeping pills in July 1979	2.1
Ratio of people using tranquilizers during the crisis to those using tranquilizers in July 1979	1.8

beverages reported an increase in their alcohol consumption during the crisis, while the comparable increase for smokers was about 32%. The total number of people using sleeping pills and tranquilizers in the two weeks following the accident was also higher than for the two weeks immediately preceding the interview in July 1979. The number of persons using sleeping pills during the two weeks immediately following the crisis was 2.1 times the number at the time of the interview. Also, the number of respondents using tranquilizers during the crisis was 1.8 times the number in July, at the time of the interview.

In the July PDH survey respondents were also presented with a list of possible coping strategies and asked which, if any, had helped them to deal with the crisis. When the responses to these questions were factor analyzed, they were found to cluster into two relatively distinct groups. (See Table 4.4.) The first consisted of a series of strategies for either distraction (concentrating on movies, TV, or reading, becoming angry, or indulging oneself) or avoidance (trying to put it out of one's mind, avoiding people, or sleeping more than

Table 4.4
Proportion of Respondents Reporting Different Coping Strategies
for Dealing with TMI Stress

Group I	
Try to forget the whole thing by going to movies, watching TV, or reading	25%
Force yourself to put it out of your mind	24%
Let off steam by getting angry	18%
Avoid people, get away by yourself	12%
Give yourself a treat by buying something you wanted	12%
Sleep more than usual	5%
Group II	
Pray for guidance	56%
Seek advice and support from friends and relatives	53%
Talk to a doctor or health professional	11%
Talk it over with a clergyman or spiritual advisor	9%
Both Groups	
Work harder at your job or around the house	28%

usual). A second group included strategies that involved a more active seeking of information and social support, including prayer and conversations with friends, relatives, health professionals, and members of the clergy. One strategy, working harder, loaded equally on both factors. Interestingly, health professionals and members of the clergy were only rarely cited as a source of support.

Characteristics of Persons Using Different Types of Coping

Analyses of who was most likely to make use of the different coping strategies showed that the choice of strategies was only weakly related to sociodemographic characteristics. Older respondents reported lower rates of psychotropic drug use during the crisis, which is consistent with the finding that older persons were less upset by the crisis initially and perceived it to be less threatening. We also found that single persons were less likely to take protective actions and less likely to engage in affiliative behavior, possibly because persons in families are more likely to be integrated into neighborhoods and hence affiliative contacts are more readily available to them. Income was related negatively to affiliation. Women were somewhat more likely to report using avoidance behaviors and tended to use more psychotropic drugs. Overall, the strength of these results was modest. Less than 10% of the variation in any of the coping responses was explained by respondents' demographic characteristics. A more detailed description of these findings is presented in Cleary and Houts (1984).

Effectiveness of Coping Resources in Moderating Stress

We next examined the extent to which coping strategies were related to the persistence of distress. Since the PDH July and January surveys interviewed 404 respondents living within five miles of TMI twice, it was possible, using multiple regression analysis, to study whether specific coping strategies were associated with change in distress levels over time. In these analyses, we examined the associations between the coping strategies reported in July and the distress (including both psychophysiological symptoms and ratings of upset) in January, while statistically controlling for symptoms and degree of upset in April. The results of these analyses are presented in Table 4.5.

Although few of the coefficients in Table 4.5 are statistically significant, there are some interesting patterns. Earlier research on stress and coping suggests that people will experience distress when they perceive a situation to be threatening and think that they lack the ability to cope. Coping may involve dealing with the threat itself (problem-focused coping) or trying to reduce the emotional impact of the threat (emotion-focused coping). It is often thought

72

Table 4.5

Effect of Coping Responses on January Distress, Controlling for April Distress (standardized regression coefficients[1])

Coping responses	Upset about TMI Beta	Psychophysiological symptoms Beta
Changed milk use	.01	.02
Scale of protective actions[2]	.12**	.01
Change in smoking	.03	.08
Change in drinking	−.01	.02
Psychotropic drug use	.02	.27**
Evacuated	.08	.00
Active in organizations and meetings	.10*	−.06
Avoidance behaviors	.09	.12**
Affiliative behaviors	.09*	.06

*p < .05.
**p < .01.

1. Each coping response was evaluated in a separate regression analysis in which the coping response and April distress were the independent variables.

2. Sum of protective actions taken: kept family indoors, sent family out of area, had tests for radiation, changed diet, and other protective actions.

that doing something (active coping) is better than doing nothing (denial) (Roth and Cohen, 1986). Thus, we predicted that taking protective actions, evacuating, being active in organizations, and being affiliated with others would lead to a reduced level of distress in January. Contrary to expectations, people who took many protective actions, who reported being active in organizations concerning TMI, and who sought out others as a means of reducing their distress were significantly more distressed in January than would be expected from the level of reported April distress. Our interpretation of these findings is that people who actively responded to the situation (e.g., by attending community meetings or talking to others about the problem) were precisely those people who were more aware of the situation and of the persistence of the threat. Furthermore, probably the most salient characteristic of the situation was that it was something about which people could do very little; therefore, going to meetings and discussing the problem with others may have increased levels of anxiety and concern. Collins, Baum, and Singer (1983) suggest a similar explanation for their finding that problem-oriented coping was associated with greater symptom reporting, while reappraisal was associated with lower symptom-reporting rates.

A useful theoretical framework for interpreting these findings is the one proposed by Seligman (1975) concerning the development of depression; if one attempts to deal with environmental events and is unsuccessful at influencing those events, one gradually develops a sense of hopelessness that becomes an integral part of the experience of depression. The circumstances at Three Mile Island did not induce widespread clinical depression, but the accident was a distressing situation over which residents in the area had no control. It is also likely that, the more they attempted to deal with the situation, the more they realized that it was beyond their control. These are the types of circumstances that sometimes lead to what Frank (1973) and Dohrenwend et al. (1980) refer to as demoralization.

When the results concerning the persistence of psychophysiological symptoms in Table 4.5 are examined, a slightly different picture emerges. None of the coping responses resulted in lower than expected rates of symptom reporting, but those persons who reported using psychotropic drugs in April and those who scored high on the scale of avoidance behaviors reported significantly more symptoms than was expected. These findings are consistent with the general literature on coping and defense. It appears that people who attempt to mask emotions (psychotropic drug use) or deny them (avoidance behaviors) are more likely to somatize their complaints and report a variety of physical complaints over a longer period. These findings are also consistent with the theoretical work of Cohen and colleagues (1985), who have argued that coping can have deleterious effects as a result of a cumulative fatigue effect, persistence of the coping activities in situations where a person is not dealing with the stressor, and side effects of the coping efforts themselves. They further argue that such effects could occur to the extent that there is a prolonged active attempt to cope, irrespective of whether the coping behavior is successful or not.

Psychological Resources

The psychological resources measured in the studies reported here are self-esteem, perceived self-efficacy, and introspection. The self-esteem scale consisted of six items derived from Rosenberg's work (1979). The scale of self-efficacy was a four-item scale constructed for this study and asking questions about whether the person typically meets problems head-on, tries to anticipate problems and make the necessary preparations, deals directly with problems and doesn't avoid them, and struggles aggressively with difficulties and doesn't give up easily. The introspection scale was based on three items: whether the respondent was sensitive and introspective, was worried about meaning in life, and was interested in psychology.

Results, shown in Table 4.6, indicate that in January 1980 lower self-esteem was significantly associated with higher reporting of psychophysiological symptoms and that introspection was related to higher levels of being upset about TMI. Since these analyses controlled for symptom and upset levels in July 1979, they indicate that low self-esteem and introspection are related to

74

Table 4.6

Relationship between Psychological Resources, Social Support, and January
Distress, Controlling for April Distress (standardized regression coefficients[1])

Independent variables	Upset about TMI Beta	Psychophysiological symptoms Beta
Psychological Resources		
Self-esteem	−.02	−.10*
Self-efficacy	.04	−.02
Introspection	.11**	.03
Social Support		
Attend church	−.03	−.04
Number of friends	−.09*	−.02
Social support	−.03	−.08

*p < .05.
**p < .01.
1. Each psychological resource and social support variable was evaluated in a separate regression analysis with April distress.

maintaining high distress levels over the six-month period between the two surveys.

The fact that introspective persons had sustained high levels of distress is counter to the common supposition that dealing directly with feelings and emotions is functional and should result in the reduction of distress. Our explanation is that, as introspective people reflected on their feelings, the incident became more salient and they were less able to ignore it. As a result, the situation continued to upset them because, as was discussed above, an important characteristic of the Three Mile Island incident was that it continued to be a potential source of distress after the initial crisis had passed. In most studies of distress, the source or cause is limited in duration. In such circumstances, it is more reasonable to expect that reflecting on the problem and "coming to terms with it" would be functional and lead to decreased distress. In the TMI situation, however, reflecting on the problem appears to have led to sustained upset. These observations are consistent with Mechanic and Hansell's (1983) argument that social and environmental discontinuities exacerbate self-awareness. Self-awareness, or introspection, is in turn related to more reporting of symptoms of distress.

The Effects of Social Support

Studies of social support tend to use two types of measures. The first type focuses on the function or quality of the relationships that a person has with other persons; the second type focuses on objective characteristics of the person's network, such as size, density, and connectedness (Cleary, 1987). In the January RWJ study, both types of measures were used, including asking respondents how many friends or relatives they felt close to, whether members of their families attended church, and four questions that formed a scale of perceived availability of social support (e.g., "Some people have difficulty finding others with whom they can discuss almost any problem they have. To what extent do you have difficulty finding such people?"). The results, shown in Table 4.6, show that there are trends for all the social resources to be associated with lower symptom reporting in January, controlling for symptom reporting the previous July, although only the relationship between number of friends and being upset about TMI was statistically significant.

The weak and inconsistent findings with respect to social support are similar to those reported by other TMI investigators. The strongest findings on social support as a buffer against stress following the accident were reported by Fleming et al. (1982), and even there the results were not consistent across different measures of distress. An accumulating body of theory and research suggests that future work should differentiate actual and perceived support, the availability and use of support, and the quantity, structure, and function of relationships (House and Kahn, 1984; Cleary, 1987; Cohen and McKay, 1984). Furthermore, it is important to make distinctions among functions such as instrumental, appraisal, and emotional support (House, 1981). Similarly, it is important to differentiate emotion management from danger control and take into account the strategies used for each (Leventhal, Safer, and Panagis, 1983).

Conclusions

In view of the intense and dramatic publicity given to the dangers at Three Mile Island during the crisis, it is not surprising that a significant number of people living near the nuclear facility experienced severe distress during and immediately following the crisis period. What was surprising was the persistence of long-term, low intensity distress, which, while it also decreased with time, was still evident in some groups of people (who were initially most distressed) for more than a year and a half after the accident. There is a consensus among all investigators that the long-term distress following the TMI accident was in the range of normal everyday stresses and far below that characteristic of mental patients.

One problem in comparing the TMI findings with those from other crises is that different measures and time frames were involved. Nonetheless, there are some studies showing long-term psychological distress following technologically caused crises (Levine, 1982; Streufert et al., 1987), and so the TMI findings on long-term stress are not unique. On the other hand, the fact that there was no physical damage to the community as a result of the TMI crisis does indicate that concern about possible harm can, by itself, cause long-term psychological distress.

Perhaps the most important difference between many of the natural disasters mentioned earlier and technological crises such as the TMI accident is that technological crises do not necessarily result in significant physical damage. If there is physical damage, there will obviously be loss-related distress. However, in disaster situations in which there is damage, people can focus their efforts on the task of recovery and repair. The process of working with neighbors on a common task that is seen as repairing the damage that was done can even have a salutary effect. In a crisis such as TMI, there was no locus of damage, the potentially harmful forces could not be seen or stopped without sophisticated technology, and people did not know how to cope with the potential threat. Perhaps the best example of this phenomenon was the picture of a woman leaving her home shortly after the crisis with an umbrella to protect herself from radiation.

The variables related to initial and subsequent distress are of two types. First are those that are related to reports of distress in other studies outside the TMI context (e.g., sex, persons with chronic conditions, and persons who tend to get upset easily) and that may partly reflect their generally higher distress levels but may also reflect higher sensitivity of those groups to stressful situations. Second are variables associated with being especially vulnerable to the dangers at Three Mile Island (e.g., living near the damaged reactor, being pregnant or having young children in the household).

Finally, an important and somewhat surprising finding was that normal methods of coping with distress were not especially effective and that active attempts to deal with the situation were associated with continuing distress. This, we argue, was because the damaged plant was seen as a continuing threat and because the persons living near it could do little to change that threat. It is likely that future crises involving complex and dangerous technologies will have similar effects, since the average person is unlikely to be able to change the situation and therefore is likely to experience persistent frustration. Further evidence of such continuing frustration is presented in Chapter 5, on attitudes concerning the TMI crisis.

5
Attitudes and Beliefs

Overview

Attitudes and beliefs concerning the crisis and its aftermath were assessed in all five telephone surveys conducted in the year and a half following the accident. Questions dealt with attitudes toward sources of information during the crisis, expectations about how the crisis would affect the area economically, attitudes concerning restarting the undamaged reactor at TMI and toward nuclear power in general, and beliefs in rumors about health effects of the accident.

During the crisis people generally relied on the media for information about the situation at TMI, though half of the respondents said they were dissatisfied with the information the media provided. Those who were upset about the accident felt that the media withheld information, while those who were not upset felt that events were blown out of proportion. Most sources of information were rated positively, but a striking 60% of respondents felt that information from Metropolitan Edison Company (the operators of Three Mile Island) had been "totally useless."

Three months after the crisis it was widely believed that the accident would lower real estate values and hurt the area economically, though subsequent objective studies showed little effect on real estate values and relatively little long-term economic impact on the area close to the reactor. Also three months after the accident, attitudes toward restarting the undamaged reactor at Three Mile Island were more negative among those living close to TMI than among those living farther away. The same was true for attitudes toward nuclear power in general. As with distress measures discussed in Chapter 4, this distance effect had largely dissipated by eighteen months postcrisis.

Eighteen months after the crisis, a majority of respondents wanted non-technically trained people to have as much influence as technicians on decisions about cleaning up the damaged reactor at Three Mile Island, suggesting a desire to introduce some external control over those who were seen as responsible for the accident. Also at this time, approximately a third of respondents believed the rumors that the accident had caused serious health and emotional problems. Interestingly, people farther from the facility were more likely to believe rumors than were people close to it, suggesting that people living close to TMI had either sought out or were exposed to factual information about the effects of the accident to a greater degree than people living farther away.

Overall, the findings indicate that people living in the vicinity of Three Mile Island formed strong opinions about and attitudes toward how the crisis was managed, about its impact on the area, and about how the continuing problems at the facility should be managed. The numbers of persons holding negative views of how the facility was managed decreased over time, though substantial numbers still expressed serious concerns eighteen months after the accident. Negative feelings about TMI were consistently associated with negative opinions about how the accident affected those living in the vicinity of the facility. Whether respondents' opinions caused the negative feelings or vice versa could not be determined.

Background

Attitudes, as traditionally defined, have two components: affective (feelings or judgments) and cognitive (opinions and beliefs). Opinions and beliefs by themselves do not include a judgment or a statement of feeling. We present results concerning attitudes and beliefs in two sections. The first section deals with beliefs, including beliefs about health-related rumors and about economic impact. The second section concerns attitudes, including attitudes toward information sources, media coverage, restart, cleanup, and nuclear power.

For each attitude or belief we first summarize related studies conducted near TMI and elsewhere by other investigators. Then we review our findings on the extent to which the attitude or belief was accepted by people living near Three Mile Island and the extent to which acceptance varied with distance from TMI, with how upset respondents were about TMI, and with whether respondents evacuated during the crisis. In the final section we examine the association between demographic characteristics such as age, sex, education, and income and attitudes toward nuclear power, both shortly after the crisis and eighteen months later.

We do not have baseline measurements of attitudes or beliefs before the crisis, and so we cannot be certain about how they changed as a result of the crisis. However, there are three ways that we can indirectly assess whether the TMI crisis affected attitudes and beliefs. First, we can examine whether they varied according to distance from TMI. We know, from findings presented in earlier chapters, that both evacuation and psychological distress were elevated near TMI, indicating that distance from the plant can be used as a proxy for degree of impact. Therefore, if attitudes or beliefs also changed with distance from TMI, this would suggest that they were affected by the crisis. The limitation of this approach is that it does not take into account how attitudes of people at all distances from TMI may have been affected by the crisis. This can be addressed by examining the relationship between beliefs/attitudes and how upset respondents were about TMI. If a belief or attitude was affected by the Three Mile Island crisis, it is likely that it is correlated with how upset a person was about the crisis, though other explanations may be possible as well. A third indicator of impact is whether beliefs or attitudes relate to behaviors such as evacuation, since people who take action in response to a situation have clearly been affected by what occurred. None of these indicators is ideal, and there are other possible explanations for the observed associations, especially for the relationships among attitudes, being upset, and evacuation. Hence, our conclusions about causes of these attitudes are tentative.

Beliefs

Health-Related Rumors

Background
Rumors, defined as "a proposition for belief in general circulation without certainty as to its truth" (Jaeger et al., 1980:473), have been the subject of extensive research by social scientists. Rosnow (1980), in a review of this literature, concludes that two conditions have been shown to affect rumor strength: apprehension about a potentially negative outcome, and uncertainty or unpredictability of what an outcome will be. Both of these conditions existed during and following the TMI crisis, as indicated earlier in this book. In addition, since the average citizen was unable to understand fully the technological aspects of the accident, uncertainty concerning what had occurred probably also facilitated the circulation of rumors.

Results
Questions about belief in rumors were included in the PDH October survey, eighteen months after the crisis. These questions concerned six rumors about purported health problems in the vicinity of TMI following the crisis: alleged increases in birth defects, stillborn children, serious emotional problems, cancer rates, general health problems, and health problems in farm animals. All of these rumors had been shown to be either incorrect or lacking objective verification. Survey responses indicate that these rumors were accepted as true by approximately a third of respondents within five miles of TMI, and rejected by about a third, with the remaining third undecided (Houts et al., 1981). For two of the rumors, increase in the number of miscarriages, stillborns, and infant deaths, and increase in birth defects, there was a statistically significant tendency for persons living fifty to fifty-five miles from the facility to accept the

rumors to a greater degree than people in the immediate vicinity of TMI. Also, people who were upset about the situation at Three Mile Island tended to believe the negative rumors more than people who were not upset (Houts et al., 1981).

Another rumor studied in the October 1980 PDH survey dealt with the danger of venting krypton gas at TMI in the spring of 1980, four months prior to the survey. Rumors about potential harm from venting were widely circulated before and during the venting, despite statements by officials that there was little danger. As with the other rumors, a substantial percent of respondents felt the venting had been dangerous (within five miles of TMI, 15% of respondents felt that the venting was very dangerous and 39% felt it was somewhat dangerous). Also, as with other rumors, acceptance of this rumor was greater fifty to fifty-five miles away than within five miles of TMI. There was also a significant correlation between this rumor and how upset respondents were about the situation at Three Mile Island (Houts et al., 1981).

Discussion

Rumors about health effects of the accident were widely shared and believed, as would be expected in a situation that was both ambiguous and anxiety provoking (Rosnow, 1980). The significant correlations between belief in rumors and degree of upset concerning TMI are also consistent with other studies of rumors discussed by Rosnow. The one surprising finding is that belief in rumors was accepted to a greater degree far from TMI than close, which is somewhat at variance with findings on stress discussed in Chapter 3. This suggests that the salience of the event to residents near the reactor led them to seek out factual information or, possibly, that such factual information was more available to them through the local media.

Economic Impact

Background

Media reports of disasters frequently include estimates of costs to local populations. There was considerable interest in the economic impact of the Three Mile Island crisis, as is evident from the studies on this subject reviewed in Chapter 3. These studies indicated that the short-term economic impact was closely linked to evacuation and that long-term impact on electricity rates was not large relative to rates of other utilities in the state. Prior to the release of those reports, however, the average citizen had little direct information with which to judge the degree of economic impact. As a result, beliefs were not constrained by facts and were open to influence by feelings and unsubstantiated rumors.

Results

Questions about perceived economic impact of the TMI crisis were included in both the NRC and PDH baseline surveys in July 1979. In the PDH survey, respondents within five miles of TMI were asked if they thought the value of their homes had changed as a result of the crisis. Thirty-two percent thought that values had decreased. This finding is identical to that reported by Barnes et al. (1979). Later studies of actual changes in real estate values in the vicinity of TMI showed no actual decrease following the crisis (see Chapter 3).

Beliefs about real estate values were found to be significantly correlated with how upset respondents were during the TMI crisis. In addition, it was found that respondents who had evacuated during the crisis were more likely to believe that real estate values had decreased than were persons who had not evacuated.

The NRC survey in July 1979 also included questions about the general economic impact of the crisis. Sixty percent of the NRC sample felt that the crisis would hurt the area economically, 34% felt that it would have no effect, and only 6% felt that the area would be helped. As with beliefs about real estate values, it was found that people living close to TMI and people who evacuated during the emergency were more likely to feel that the crisis would hurt the area economically (Flynn, 1979).

Discussion

The findings from the PDH and NRC surveys are consistent in that they both indicate that significant percentages of the population near TMI felt that the crisis would have negative economic effects on the area. The fact that objective studies showed limited long-term economic effects suggests that these beliefs were influenced by people's feelings about TMI. Alternatively, it could be that people were more upset because of the perceived impact on the value of their property.

Attitudes

Information Sources

Background

Research on attitude change indicates that the prestige of the information source plays an important role in how people respond to the information provided (e.g., Andreoli and Worchel, 1978; Babad, 1977). In the Three Mile Island situation, people had to weigh the prestige of Metropolitan Edison (Met Ed) against that of government officials and agencies, since these sources were often in conflict. For example, Met Ed downplayed the seriousness of the accident, while other sources stated that there was a threat to public safety.

National surveys conducted by Gallup and CBS after the TMI crisis (Nealy et al., 1983) showed that a majority of respondents felt negatively about Met Ed as a source of information, while a majority were positive about the government officials and agencies involved. The Nuclear Regulatory Commission, on the other hand, received the lowest ratings of the government agencies involved. The studies reviewed here have investigated these same attitudes among persons in the immediate vicinity of Three Mile Island.

Results
The NRC survey in July 1979 asked respondents how they felt about several sources of information during the two-week crisis. The governor of Pennsylvania and the Nuclear Regulatory Commission were rated most positively. Intermediate ratings were given to state emergency agencies, local government agencies, and the president of the United States. Markedly lower ratings were given to Metropolitan Edison, with 60% of respondents feeling that the company was "totally useless" (Flynn, 1979). Sorensen et al. (1987) report that the low ratings of Metropolitan Edison as a source of information were repeated in two later surveys of TMI area residents conducted for the Nuclear Regulatory Commission in June 1980 and in March 1981.

Analyses of the Flynn data show no statistically significant differences between ratings of sources of information by people who evacuated during the crisis and by those who did not; however, respondents who were upset about TMI were more likely to rate Met Ed unfavorably than people who were not upset. There was also a tendency for people close to the TMI facility to rate information from the NRC and the governor of Pennsylvania as extremely useful and to rate information from Met Ed as useless.

Discussion
The most striking finding regarding sources of information was that approximately two out of three respondents felt that Met Ed was "totally useless" as a source of information. Other sources were generally rated positively. The poor ratings given to Met Ed probably result from the fact that the company's news releases during the crisis often contradicted what other groups said; consequently, people had to choose between Met Ed and sources of information that were either more prestigious or were seen as more objective. Negative attitudes toward Met Ed as a source of information were related to having evacuated during the crisis, shorter distance from TMI, and being upset about TMI, suggesting that people who felt more threatened by the situation were more likely to be angry toward those seen as responsible for the threat. The main difference between the findings reported here and those reported in national surveys is the high support for the NRC among persons close to TMI.

Media Coverage

Background

There have been several articles on media coverage of the TMI crisis that discuss the difficulties reporters faced in obtaining consistent and understandable information about what was occurring (e.g., Burnham, 1981; Nimmo, 1984; Rubin, 1981; Sandman and Paden, 1979). These articles indicate that, even among professional journalists, there was disagreement about how objective and fair the reporting was during the crisis. Earlier research on how the public views the media indicates that confidence in the media had decreased from the 1950s to the time of the TMI accident. A CBS national poll conducted following the TMI crisis found that 57% of respondents believed that events during the crisis were reported fairly, while 28% felt that the situation was blown out of proportion. They also found that 52% of pro-nuclear people felt the reporting was fair, while among anti-nuclear respondents 64% thought it was fair (Mitchell, 1980; Nealy et al., 1983).

Questions about how people felt about media coverage of Three Mile Island were asked twice: in the July 1979 NRC survey, and fifteen months later in the October 1980 PDH study. The NRC questions dealt with the usefulness of different ways of obtaining information and whether respondents were satisfied with the information they received during the emergency. The October 1980 survey dealt with whether respondents felt that the media withheld information, reported it accurately, or blew events out of proportion.

Results

The NRC survey showed that two-thirds of respondents made their decisions to evacuate based on TV or radio reports, while only 11% based their decisions on what they heard from friends. As a result, respondents gave high ratings for usefulness of local TV and radio during the crisis, but sharply lower ratings for the usefulness of information from friends and relatives. Intermediate were national network TV, newspapers, and national news magazines. Flynn (1979) suggested that the reason that friends and relatives were rated as poor sources of information is that they were seen as spreading rumors, while the media was seen as providing factual information. Flynn also reported that neither distance from TMI nor whether the respondent evacuated during the crisis was related to ratings of different media (Flynn, 1979).

Ratings of satisfaction with the media in the NRC survey were approximately evenly split, with half the respondents saying that they were satisfied and half dissatisfied with the information they received. Satisfaction with the media was related to distance from TMI, with persons living closer to TMI reporting that they were more dissatisfied. Also, people who evacuated during the crisis reported being less satisfied with media coverage of TMI-related events than did people who did not evacuate (Flynn, 1979).

84

Fifteen months after the NRC survey, the PDH October 1980 survey showed that less than a third of respondents felt that events were reported accurately, while approximately half of the respondents felt that the media had blown events out of proportion and 22% felt that information was withheld or covered up. These percentages were approximately the same for persons living at different distances from the facility. However, there were sharp differences between people who were and were not upset about the situation at TMI. Respondents who were upset tended to think that information was withheld, and people who were not upset tended to think that events were blown out of proportion by the media. Respondents who had evacuated during the crisis also tended to think that information was withheld. Since a national study had shown a relationship between attitudes toward nuclear energy and satisfaction with media coverage at TMI (Nealy et al., 1983), we repeated this analysis for the population near TMI. The results with the TMI sample showed the same relationship, with pro-nuclear respondents feeling that events were blown out of proportion and anti-nuclear respondents tending to feel that information was withheld.

Discussion

Overall, these results indicate that, while people at all distances from the facility relied primarily on the media rather than person-to-person interactions for information about TMI, there was considerable dissatisfaction with the coverage provided. Dissatisfaction near TMI was greater than the CBS poll found for the nation as a whole. The findings near TMI indicated that respondents' attitudes toward the media were associated with their general attitudes toward TMI. People who were not upset about the accident felt that events were blown out of proportion, and people who were upset felt that information was withheld. This suggests that people's judgments about the media were partly based on whether they agreed with what the media reported.

Restarting the Undamaged Reactor at TMI

Background

The accident at Three Mile Island occurred in reactor #2. At the time of the accident, reactor #1 was temporarily shut down for routine maintenance. Since the two reactors function independently of one another, reactor #1 remained unaffected by the accident at reactor #2. Shortly after the crisis, Met Ed asked the NRC for permission to restart reactor #1. This request became the center of a public controversy that lasted a number of years, until reactor #1 was restarted in October 1985. During this time the issue became a test of confidence in Met Ed's performance during and after the crisis. Debate was further intensified by allegations of Met Ed employees' cheating on NRC qualification examinations and of the covering up of data on the malfunctioning of reactor #2 prior to the accident. Walsh (1986) has discussed the role that these allegations played in efforts to delay the restart of TMI's undamaged reactor. Questions concerning attitudes toward restart were included in the two PDH surveys conducted in January and October 1980, nine and eighteen months following the crisis.

Results

The January 1980 PDH survey, which included persons living as far as fifty-five miles from TMI, asked respondents how they felt about restarting TMI. The options were: support restart, don't care one way or the other, or against restarting it. Sixty percent of respondents within five miles of TMI were against restart, while approximately 50% were opposed 6–15 miles away, 40% were opposed from 16–25 miles away, and 30% were opposed beyond 25 miles. This strong relationship between distance from TMI and support for restart is statistically significant (Houts and Goldhaber, 1981). Not surprisingly, people who said that they were upset about TMI were significantly more likely to oppose restart than were people who were not upset.

The October 1980 PDH survey, which included just persons within five miles of TMI and a control group 41–55 miles away, asked a slightly different question: "How do you feel about restarting TMI unit one, the reactor not involved in last year's accident?" Options were the same as in the January survey. Results showed a drop in percent opposing restart within five miles from 60% to 46%, but an increase in percent opposing restart 41–55 miles away, from 30% to 42%. The drop for persons opposing restart within five miles was a statistically significant change, while the increase in the 41–55 mile group was not (Houts and Goldhaber, 1981). While these findings suggest a decrease in opposition to restart over time, it should also be noted that the phrasing of the question was different in the two surveys. This could have additionally influenced responses. As in the January survey, there was a strong, statistically significant association between being upset about TMI and being opposed to restart.

In the January survey we examined how attitudes toward restart related to two behaviors: evacuation during the crisis, and participation in political activity concerning TMI. Evacuation was not significantly related to restart attitudes, though the trend was for evacuees to oppose restart. However, opposition to restart was associated with greater political activity. In October both evacuation and political activity were significantly related to opposition to restart. Additionally, it was found that evacuation during the krypton venting, in the spring of 1980, was significantly related to opposition to restart. One interesting difference between the January and October findings is that in January 100% of those who took part in political activities regarding TMI were opposed to restart, while in October only 71% of the political activists were opposed to restart. It is likely that this change reflects the fact that, as pointed out by Walsh (1983), a citizen group supporting TMI did not get underway until a year after the accident. Therefore, people supportive of restart became politically involved later than did persons who opposed restart.

Discussion

These findings show that opposition to restarting the undamaged reactor at TMI followed a pattern similar to that found for distress: higher opposition close to TMI than farther away, but with less of an association with distance as time passed. Opposition to restart was also strongly associated with how upset respondents were about TMI and with their evacuation and political behaviors following the accident. The association between attitudes and evacuation both at three and eighteen months suggests that they were stable over the fifteen-month period. Dew et al. (in press) have shown that attitudes toward a variety of issues at TMI were very stable over a thirty-one-month period, from nine to forty-two months after the accident.

Several other groups of investigators also reported findings on opposition to restart. Sorensen et al. (1987) report 46% disapproval of restart within five miles of TMI in June 1980 and a 40% disapproval rate in March 1981. Their findings, which are in substantial agreement with the October PDH results, were based on two random digit dialing surveys involving 623 interviews within five miles of TMI in June 1980 and 440 interviews in March 1981. However, another group of investigators (Schorr et al., 1982), who obtained data on attitudes toward restart in October 1980, reported a higher rate of opposition (72%) than was found in the PDH survey. A careful examination of that study suggests two possible reasons for the difference in findings. First, the Schorr et al. study was conducted just in Newberry Township and Goldsboro, which include only part of the population within five-miles of TMI, while the PDH survey sampled the entire five-mile area. Our data do not allow us to separate this area from other areas within the five mile radius, but it is possible that opposition was greater there than in other areas within five miles of TMI. A second explanation is that the Schorr et al. sample included only persons who had responded to three waves of interviews over a twelve-month period. It is possible that persons with strong negative feelings about TMI might be especially motivated to respond to three separate interviews. Furthermore, the Schorr et al. sample was weighted toward females (70% of their sample), who were found, in our surveys, to be especially likely to be opposed to restart (Houts et al., 1983).

Interestingly, two surveys by Kraybill (1979; 1980) also appear to disagree with the PDH survey findings, but in the opposite direction. The first Kraybill survey, which was conducted approximately a week after the accident, found that only 40% of respondents within five miles of TMI wanted the plant closed. The second survey, which reinterviewed 74% of those respondents one year later in March 1980, found that only 44% wanted the plant closed. This contrasts with our finding nine months after the accident that 60% of respondents within five miles of TMI opposed restart. While there were some differences in how the questions were phrased, which could account for some of the differences, it should also be noted that the Kraybill studies undersampled people who evacuated during the crisis, since his first sample was conducted while many evacuees were still away. This sample bias may have also affected his findings. (See Chapter 2 for a discussion of possible sample biases in the Kraybill study.)

Cleaning Up the Damaged TMI Reactor

Background
Responses to questions about who should influence the cleanup at TMI provide insights into people's attitudes about who should control technology. Attitudes toward technology have been widely studied by public-opinion pollsters. These studies have consistently reported positive attitudes on the part of the American public toward science and technology, and while some studies have indicated a decrease in support for science and technology in the years just prior to the crisis, overall levels of support have remained high (Marshall, 1979; Mazur, 1981; Nesterenko, 1984). At the same time, it should also be noted that there is evidence that many people feel ambivalent about technology, since in several studies the same respondents who support technology also express concerns about the impact of technology on society (Marshall, 1979; Nesterenko, 1984). The accident at Three Mile Island provided an opportunity to study how exposure to a technologically caused crisis affects attitudes toward controlling technology.

Shortly after the crisis, politicians at the local, state, or federal levels, as well as spokespersons for various interest groups, indicated that they wanted a role in decisions about the future of TMI. Questions about how respondents felt about different groups' participating in TMI decisions were included in the PDH survey of October 1980, eighteen months after the crisis. In that survey, respondents within five miles of TMI and a control group forty to fifty-five miles away were asked how much influence each of the following groups should have in determining how the cleanup was carried out: 1) organized groups of citizens; 2) energy experts from colleges, universities, and other impartial organizations; 3) the general public; 4) elected local officials; 5) Metropolitan Edison Company; 6) Nuclear Regulatory Commission; 7) energy experts working for the Pennsylvania Department of Environmental Resources; and 8) elected state or federal officials. The response options were none, some, or very much influence.

In analyzing responses to these questions, we first determined which groups received the most and the least support. Next we divided the interest groups into two categories, groups with and without technical expertise, and examined the support given to each category.

Results

Results indicate relatively high support for all eight groups. Even organized groups of citizens, the group with the lowest level of support, received some support from over 70% of respondents. In general, persons living close to TMI and those living far away rated the groups similarly, with the NRC and the Pennsylvania Department of Environmental Resources (DER) receiving the highest ratings (Houts et al., 1981). There were statistically significant relationships between how upset respondents were about TMI and the degree of influence they believed that all groups, except DER, should have in the cleanup. For Met Ed and the NRC, the direction of the relationship was negative, indicating that people who were upset about TMI wanted those groups to have less influence. Other relationships were positive, with upset people wanting groups to have greater influence.

To explore whether events at Three Mile Island had affected people's attitudes toward control of technology, each respondent was assigned a score representing whether he or she wanted technicians (NRC, DER, Met Ed, or energy experts from impartial organizations) to have more, the same, or less influence than nontechnicians (politicians, organized groups of citizens, or the general public). This score was computed by comparing the highest score given any technician group with the highest score given any nontechnician group. The results indicate that only 8% wanted nontechnicians to have greater influence than technicians, 36% wanted technicians to have more influence than nontechnicians, and 56%, the largest group, wanted equal influence by technician and nontechnician groups. Scores on this scale were significantly related to how upset respondents were about TMI. People who were upset tended to want nontechnicians to have more say than technicians, suggesting some degree of anti-technology attitudes among those most upset by the crisis.

Discussion

The findings concerning who should influence decisions at Three Mile Island indicate that there was dissatisfaction with the Metropolitan Edison Company and with the Nuclear Regulatory Commission as the sole decision makers about the future of TMI. This is consistent with the negative attitudes toward Met Ed as a source of information discussed earlier; however, it is not entirely consistent with the high ratings given to the NRC as a source of information shortly after the crisis. Since these ratings about participation in decisions were obtained eighteen months after the crisis, this suggests a change in attitudes toward the NRC over the fifteen-month period between the two surveys. The results also showed widespread support for groups without technical expertise participating in the decision-making process. While only a small minority wanted nontechnicians to have more influence than technicians, a sizeable majority wanted nontechnical groups to have at least as much influence as the technical ones. This does not so much suggest being disillusioned by technology as a desire for a sharing of power with technicians in controlling technology.

In reviewing other studies of attitudes toward technology, we noted that, while most people support technology, many also feel concern about its impact on society. The findings reported here are similar in that they indicate that most respondents, while wanting technicians to play a major role in decision making, did not trust them to make the decisions alone. They wanted nontechnicians to play a role also.

Nuclear Power

Background
Since Three Mile Island is a nuclear power plant, it is likely that general attitudes toward nuclear power were also influenced by the crisis at TMI. National public-opinion polls by Harris showed a decline in support for construction of nuclear power plants even before the TMI crisis (Firebaugh, 1981; Marshall, 1979). Van der Pligt et al. (1986a) also reported widespread opposition to building nuclear power stations in England. Mitchell (1982) reported that a national poll following the TMI crisis showed an increase in the percent of persons who felt both strongly for and against nuclear power, suggesting that the crisis may have caused an increased polarization of views. Nealy et al. (1983) have also reported that persons living near nuclear power plants tend to be more supportive of nuclear power than persons living farther away. However, Fruedenberg and Baxter (1984) have shown that, since the TMI accident, opposition to nuclear plants in communities where they are either planned or being built has increased to the point where a majority oppose construction. Van der Pligt et al. (1986b), on the other hand, found that persons living near existing nuclear power plants in England were more favorable toward building additional plants than were those living farther away.

One other TMI investigator asked about attitudes toward nuclear power. Kraybill (1980) reported that a survey of people within five miles of TMI approximately a week after the accident found that 57% said they supported the use of nuclear power as a source of energy for the nation. Reinterviews with 74% of those respondents in March 1980, one year later, showed a drop to 53% supporting nuclear power.

Questions about attitudes toward nuclear power were asked twice: in the NRC July 1979 survey and eighteen months later in the PDH October 1980 survey. The NRC July 1979 survey asked respondents to weigh the advantages and disadvantages of both nuclear power in general and of the TMI facility in particular. The PDH October 1980 survey asked if respondents felt that, in the future, this country should have more nuclear plants, the same number as now, or fewer nuclear plants.

Results
The NRC survey found that approximately a third of the respondents felt that the advantages of nuclear power outweighed the disadvantages, a third felt that they were the same, and a third thought that the disadvantages outweighed the advantages. Persons living within fifteen miles of the facility were more likely to feel that disadvantages were greater, while persons beyond fifteen miles were more likely to feel that advantages were greater. Respondents who had evacuated during the crisis and who were upset about TMI were more likely to feel that the disadvantages of nuclear power outweighed the advantages.

Responses to the question about advantages versus disadvantages of the Three Mile Island facility were more negative, with half of respondents feeling that the disadvantages outweighed the advantages and only 17% saying that the advantages outweighed the disadvantages. As with the ratings of nuclear power in general, more respondents close to TMI than farther away felt that the disadvantages outweighed the advantages. The NRC survey also found that people who evacuated during the crisis and persons who were upset about TMI were more likely than others to be negative about nuclear power in general and about TMI in particular.

In October 1980 the distribution of views about nuclear power was similar to those reported in the NRC survey fifteen months earlier. Thirty-eight percent of respondents said that they wanted more nuclear plants in the future, 32% said they wanted fewer, and 30% wanted the same number as now. Persons who had evacuated during the crisis and persons who said they were upset about TMI also tended to want fewer nuclear plants in the future. However, the association between distance from TMI and attitudes toward nuclear power was no longer statistically significant as it had been in the NRC survey fifteen months earlier.

Discussion
It is difficult to compare the results reported here with those of the Kraybill surveys because the questions about nuclear power were phrased differently and the samples were obtained differently, as explained in Chapter 2. The findings that distance from TMI, evacuation during the crisis, and being upset about TMI are all related to attitudes toward nuclear power shortly after the crisis suggest that the TMI crisis did have an impact on people's attitudes toward nuclear power. This is especially clear when comparing the NRC findings with those of Nealy et al. (1983), who reported more support for nuclear power among persons living near nuclear power plants prior to the TMI crisis, the reverse of the NRC study findings. This pattern of local opposition to

nuclear power after the TMI crisis extended to other sites as well, as shown by Fruedenberg and Baxter (1984). The finding, in the NRC survey, that attitudes toward TMI were different from attitudes toward nuclear power in general suggests that many respondents distinguished between events at Three Mile Island and the industry as a whole. A likely reason is that Met Ed took much of the blame for the accident, indicated by the very low credibility ratings the company received, thus enabling many people to view other aspects of the nuclear industry differently.

Characteristics of Respondents Holding Different Attitudes toward Nuclear Power

Background
Since many attitudes are analyzed in this chapter, it is not feasible to report demographic characteristics of persons agreeing or disagreeing with each attitude. Therefore, we have restricted our discussion to characteristics of persons who are opposed to or supportive of nuclear power. This attitude was chosen because it was assessed twice, in July 1979 and in October 1980, allowing us to examine whether there were changes over time. Furthermore, there have been other studies of characteristics of persons opposing nuclear power to which we can relate our results.

A review of research on attitudes toward nuclear power by Nealy et al. (1983) provides useful background information against which to compare our findings. Nealy et al. reported that, prior to the TMI crisis, a higher percent of women than men consistently opposed nuclear power. Furthermore, national studies following the TMI crisis also showed that women's attitudes became more dramatically anti-nuclear than did men's. Nealy et al. reported that the relationships between education and attitudes toward nuclear power have not been as consistent as for sex, but some investigators have reported a tendency for lower-educated persons to be less supportive of nuclear power than persons with higher levels of education. Results with income are similar, with some investigators showing a tendency for lower-income persons to oppose construction of nuclear plants. Age has been reported to have a curvilinear relationship with attitudes toward nuclear power, with opposition highest among the youngest and oldest respondents. Finally, national polls have shown higher opposition to nuclear power in the northeastern states, where TMI is located.

Results
The NRC survey, three months after the crisis, found that women and persons with lower incomes tended to feel that the disadvantages of nuclear power were greater than the advantages. However, there were no statistically significant

92

associations between feelings about nuclear power and age or education, as in the national studies cited above. We also tested for possible curvilinear relationships with age, but none was detected.

In the October 1980 PDH survey, education, sex, and income were all significantly associated with attitudes toward nuclear power. As in the studies cited earlier, lower-educated respondents, women, and lower-income persons tended to oppose nuclear power. As in the NRC survey, age was not significantly correlated with attitudes toward nuclear power. We investigated whether the relationship between age and attitudes toward nuclear power was curvilinear, and again found no such pattern in our data.

Discussion

The relationships between both sex and income and attitudes toward nuclear power among persons living in the vicinity of the reactor were similar to those reported in surveys of other populations prior to the accident. This indicates that the TMI crisis did not change those associations, even among the persons most directly affected by the crisis. There was an association between education and attitudes toward nuclear power in the October survey, which is consistent with findings by other investigators. The lack of any significant relationship between age and attitudes toward nuclear power in either of the surveys suggests that this association may be less stable than implied by national surveys.

Conclusions

Feeling upset about Three Mile Island was associated with many of the beliefs and attitudes discussed in this chapter, including believing health-related rumors, believing that the area would be hurt economically by the crisis, attitudes toward sources of information during the crisis and toward media coverage of the crisis, as well as attitudes toward restart, nuclear power, and who should have a say in the cleanup of the damaged reactor at Three Mile Island. This suggests that many people developed a positive or negative TMI mind-set and that attitudes toward specific issues were generally consistent with that mind-set. The fact that belief in health-related rumors and beliefs about adverse economic impact of the crisis were not supported by objective data is a further indication that people's general mind-set toward TMI may have affected how they interpreted information about TMI. One explanation for these findings is that people who were more upset by the incident and had negative attitudes were more likely to be aware of and believe negative rumors. However, an equally plausible explanation is that people who had less accurate information were more likely to be upset and/or have negative attitudes. Unfortunately, the available data do not allow us to test which explanation is correct.

The clearest evidence for the impact of the TMI crisis on beliefs and attitudes is the tendency for people who live close to the facility to have more negative attitudes than persons living farther away, a tendency that parallels the findings for distress discussed earlier in this book. People living closer to TMI were found to have more negative attitudes about TMI in the July 1979 and January 1980 surveys. However, by October 1980, this distance effect had largely disappeared, and we even found that people living farther from TMI tended to believe some rumors to a greater degree than did those living close by. We speculate that this was because people living near TMI were exposed to or sought out factual information to a greater degree than did people living farther away. These findings of a diminution of negative attitudes toward TMI as time passed appear, at first, to contradict Dew et al.'s findings of relatively little change in attitudes toward TMI over a thirty-three-month period following the accident. Our explanation is the same as proposed in the previous chapter on distress, namely, that the Dew et al. sample consists of mothers of young children and workers at the TMI facility, who are among those most negative and most positive toward the facility. The high degree of persistence of their attitudes is because of their special concerns and commitments, which are not characteristic of the population as a whole near the facility.

In general, attitudes among persons close to TMI were similar to those reported in surveys of similar topics with other populations. This includes national polls on attitudes toward nuclear energy, technology, and acceptance of rumors. However, in the case of attitudes toward nuclear power, where other investigators have reported an association between age and attitudes toward nuclear power, we did not. Attitudes concerning who should participate in decisions about the cleanup indicated the degree to which the Three Mile Island experience affected general attitudes toward technology. We found that a majority of people wanted nontechnicians to have an equal role with technicians in planning how to conduct the cleanup of the damaged reactor and that people who were upset about TMI were especially likely to want nontechnicians to have a say in decisions concerning the future of TMI.

6

Conclusions

In this chapter we summarize the findings on the scope and persistence of the effects of the TMI crisis on the people living in the vicinity. We also discuss how three aspects of the crisis (technological complexity of nuclear power plants, radiation as a source of danger, and attributing blame to those viewed as responsible for the accident) played a central role in how the crisis affected the surrounding population. We conclude with a discussion of the implications of these findings for planners of emergency services and social scientists.

Scope and Persistence of the Impact of the TMI Crisis: Summary of Findings

A consistent pattern in the diverse findings reviewed in this book is that, during a period of one to two weeks immediately following the accident, there were substantial psychological, social, and economic effects on the population living in the vicinity of Three Mile Island. Long-term effects, on the other hand, varied in length and severity with the type of impact studied.

Short-term evacuation was substantial, with an estimated 144,000 people who lived within a fifteen-mile radius of the facility leaving their homes. The average time away was between four and five days, with most people staying with relatives or friends. Sixty-six percent of households within five miles had at least one person evacuate, with the percent of evacuees dropping sharply with distance from the plant to one percent beyond forty miles from TMI. Households that evacuated tended to have characteristics similar to persons who were advised to evacuate, that is, they were located closer to TMI and included young children. Long-term evacuation rates (i.e., rates of persons permanently moving out of the area), on the other hand, were not affected by the crisis.

Short-term costs of the TMI crisis were much lower than for most natural disasters, primarily because the accident caused no physical damage, contrary to what occurs in floods or hurricanes, for example. Short-term costs consisted primarily of expenses borne by families that evacuated and by businesses because of lost sales and production. Businesses had the greatest costs, while evacuation costs were minimized because most evacuees stayed with family or friends. There was little evidence of the accident's long-term economic impact on persons living in the vicinity of Three Mile Island. For example, the crisis did not affect real estate values near the facility.

The costs of cleaning up the damaged reactor (estimated at more than a billion dollars) and of purchasing replacement power to make up for what TMI would have produced were spread widely. Stockholders of GPU (the company that owned Three Mile Island) sustained substantial costs in the form of lost dividends, estimated at more than $800 million over the eight-year period following the accident. Insurance companies, federal and state governments, and the customers and stockholders of electric power companies throughout the nation also contributed substantially to defray these costs. In addition to contributing to the cleanup, governmental and industry groups took a number of other steps to shelter the company and its customers from the financial impact of the accident, including selling replacement power at reduced prices, reducing taxes on purchases of replacement power, and allowing accelerated amortization of the damaged reactor.

The reporting of stress-related symptoms among people living near TMI was most intense during and immediately following the crisis period but was of much lower intensity after that. It was estimated that, during the crisis, there was an increase of approximately 10% in the number of persons reporting symptoms characteristic of mental patients, and studies of mothers of young children living near TMI found increased frequency of clinical episodes of anxiety and depression during the period immediately following the accident, decreasing over the following year. Surveys three and nine months after the accident still showed approximately a 15% elevation near TMI in the number of people reporting stress-related symptoms compared to forty to fifty-five miles away, but the intensity of those symptoms was in the range of everyday stressors and far below those characteristic of mental patients. Other investigators also found evidence for the crisis's long-term effects on symptom reporting and agree that these long-term effects were of low intensity. The heightened symptom reporting near TMI gradually decreased with time, until in October 1980, eighteen months after the crisis, it was no longer significantly higher near the facility in the general population studies. However, other investigators' studies of subpopulations that were among the most distressed by the crisis suggest that some people in those groups continued to experience low-level distress beyond the eighteen-month period for as long as six years.

Care should be taken in interpreting the higher symptom-reporting rates near TMI because, as explained in Chapter 4, we cannot differentiate, from survey data, people who actually experienced more physical symptoms because of the TMI crisis from those who remembered or noticed symptoms due to other causes. While some investigators reported physiological evidence for increased stress, there is little evidence that the stress levels affected utilization of physicians' services or resulted in physician-diagnosed illnesses. Epidemiologic studies are being conducted by the Pennsylvania Department of Health and other groups to assess, as objectively as possible, whether the radiation released during the crisis or stress experienced during or after the crisis affected the health of persons living near Three Mile Island. These studies will continue for many years. As of the publication of this book, no definitive epidemiologic evidence of physical health effects attributable to living in the vicinity of Three Mile Island at the time of the nuclear accident has been reported.

Attitudes toward TMI and related issues (e.g., nuclear power, media coverage, reopening the facility) followed a pattern similar to that found for symptom reporting. Just after the crisis, attitudes were markedly more negative among persons living near TMI as compared to those living forty to fifty-five miles away. However, eighteen months later most attitudes were no longer significantly different for persons close to versus far from TMI, and in the case of rumors, we found that some negative rumors were accepted to a greater degree far from the facility than close by. Even though the differential between those close to and far from TMI decreased with time, there were still substantial percentages of people both close to and far from TMI who reported serious concerns about Three Mile Island eighteen months after the accident. For example, over 40% of respondents both close to and far from TMI eighteen months after the crisis still opposed restarting the undamaged reactor, and approximately 15% of both groups still reported being very or somewhat upset about the situation at Three Mile Island.

A pattern that was repeated several times in the findings reported in this book was a discrepancy between people's perceptions of what occurred and what actually happened. These discrepancies were all in the direction of people believing that the crisis had a greater impact than it actually did. In four areas data on subjective responses could be compared to objective data. The first was belief in rumors about alleged, but unproven, health effects of the accident, where it was found, eighteen months after the crisis, that approximately a third of respondents believed these rumors and 17% also said that they had firsthand knowledge to support their beliefs. The second concerned real estate values, where it was found, in July 1979, that 32% of respondents thought that the value of their homes had decreased and 9% said they knew people who had received less money for their homes because of the TMI accident. However, subsequent studies of real estate sold near TMI showed no significant change in either volume or value of sales in the year following the crisis.

The third discrepancy concerned mobility, that is, people changing residences following the crisis. In interviews three months after the crisis, 6% of respondents within five miles of TMI said they had decided to move because of the accident, and interviews with people who did move after the crisis found that 19% cited the TMI accident as a major cause of their changing residences. However, objective studies of mobility rates within five miles of TMI indicated no difference from a control group or from what would be expected based on past mobility in the area. The fourth discrepancy was in use of physician services. Nine months after the crisis, 11% of respondents within five miles of TMI stated that they had visited a physician concerning symptoms they thought were due to the situation at Three Mile Island. However, a survey of physicians a year after the crisis indicated that, in the physicians' judgments, only 25% of patients who felt their symptoms were due to TMI were, in fact, experiencing symptoms caused by stress resulting from the crisis. In addition, Blue Cross/ Blue Shield claims for physicians practicing in the immediate vicinity of Three Mile Island showed little, if any, increase in claim rates during the year following the crisis. For all four of these discrepancies, people who were more upset about the crisis were more likely than those not upset to overstate the impact of the crisis. Whether people's misperceptions made them upset or whether being upset made them misperceive cannot be determined from the available data.

Finally, it should be noted that the statistical analyses discussed here explain only a small part of the variation in attitudes, emotions, and behaviors studied. This means that there were many factors, in addition to those reported here, that affected people's reactions to the crisis. While the statistically significant findings extend our understanding of why people responded as they did to the crisis, the large amount of unexplained variance tells us that this understanding is limited and that there is much work still to be done.

Characteristics of the TMI Crisis That Affected Its Impact

In the preface to this book we proposed three features of the Three Mile Island crisis that were central in determining how the crisis affected people living in the vicinity of TMI. The first of these was the complexity inherent in nuclear power technology, which resulted in high costs for repairing the damage to the facility. The technological complexity also made it very difficult for the general public to understand technical problems when they occurred and made them dependent on technicians, who did not always agree in judging the degree to which the public was in danger. The second feature was radiation, which is an especially frightening source of danger because of its association with atomic bombs and cancer, its capability of spreading quickly over a large geographic area, the impossibility of sensing, without instruments, when one is being irradiated, and the fact that health effects of radiation from TMI would not be known for many years after the accident. The third feature was that, in contrast to natural disasters, where people usually accept the danger as inevitable or at least beyond their ability to control, at Three Mile Island, as in all technologically caused crises, it was possible to assign blame to persons seen as responsible for the danger.

98

These three features of the TMI crisis contributed significantly to many of the impacts discussed in this book, especially where TMI differs from what had been reported for natural disasters. For example, in Chapter 2 it was noted that evacuation during the TMI crisis was unusual in two ways: 1) many more people evacuated than were advised to do so, which is the opposite of what occurs in most natural disasters, and 2) disagreement among experts was a commonly cited reason for evacuating, while, in natural disasters, disagreements are often cited as a reason for not evacuating. Fear of radiation contributed importantly to the extent of evacuation because, if radiation were to escape from the reactor building, a large geographic area would be affected. Furthermore, the fright that radiation elicits and the inability to sense when it is a danger, as pointed out above, probably also contributed to the scope of the evacuation. The complexity of nuclear technology affected evacuation because it made the public dependent on experts to evaluate the danger. Therefore, when the experts disagreed, the level of uncertainty and concern among the public increased. In natural disasters, where people often feel they are able to judge the danger themselves, disagreement among experts leads people to make their own estimates, which often leads to underestimating the true danger.

The crisis's economic impact was also influenced by these three features. An important and unusual aspect of the Three Mile Island crisis was that long-term economic impact for the nation as a whole greatly exceeded the short-term impact to the immediate area where the accident occurred. In many, if not most, natural disasters, the physical damage is immediate and localized and most of the costs are incurred shortly after the disaster in repairing that damage. In the nuclear accident at Three Mile Island, radiation, which was the major source of danger, did not cause any physical damage to the surrounding community, and so most of the short-term costs were related to evacuations during the crisis. Long-term costs of over a billion dollars were primarily to clean up the damaged reactor, and they were not especially borne by people living near the facility. Instead, these costs were absorbed by stockholders, governments, and purchasers of electricity throughout the country. The high costs of cleanup were largely due to the complex and expensive technology of nuclear power generation and the dangers in handling radioactive materials. In addition, the blame directed at the company operating the facility led to political and legal efforts to prevent restarting TMI's undamaged reactor. While these efforts did not directly cause delays in restart, they contributed to nationwide concerns about nuclear power, which, in turn, increased pressures on the Nuclear Regulatory Commission to become more rigorous in monitoring construction and operation of all nuclear power plants, including the TMI facility.

Psychological distress was affected by both the complexity of the technology and by the special characteristics of radiation as a source of danger. The complex technology had psychological effects because it made the public dependent on experts to evaluate the degree of danger and to tell them how to protect themselves. Therefore, when the experts were not trusted or when they disagreed with each other, as happened during the TMI crisis, this was especially upsetting to people who depended on them for their safety. Following the crisis, many persons were apprehensive about future problems at the plant because they had lost trust in those responsible for its operation. The complexity of the nuclear power plant technology also contributed to the frustration of persons living near the facility following the crisis, since it limited the ability of laypersons to influence decisions about how the cleanup was conducted and to affect the planning of how to prevent future problems at the facility. This frustration was evident in the persistence of distress, especially among those who used active coping strategies, such as attending meetings or talking to people. Concern about the long-term health effects of radiation contributed to maintaining distress among persons living near TMI, especially mothers of young children or women who were pregnant during the crisis, because of the vulnerability of fetuses and young children to radiation.

The findings on attitudes following the crisis clearly reflected the tendency to direct blame toward those perceived as responsible for the accident. One of the most striking findings on attitudes was the overwhelmingly negative feelings about Metropolitan Edison Company (which was operating the plant at the time of the accident for its parent company, General Public Utilities, the owners of Three Mile Island), which was commonly blamed for the crisis. Sixty percent of respondents to a survey three months after the accident rated the information from Met Ed as "totally useless." Interestingly, attitudes toward nuclear power and other related subjects were not as negative, indicating that the blame was focused on those responsible for operating the facility and did not generalize widely. A second interesting finding was that a majority of people near TMI wanted nontechnicians to have at least as much influence in how the cleanup was conducted as the technicians. This could be interpreted as a desire to gain some control over technicians, in whom they had lost trust.

Implications of the TMI Crisis for Planners and Social Scientists

Implications for Planners

Several aspects of the Three Mile Island crisis have implications for public officials responsible for planning how to deal with public crises. First, in a technologically caused crisis, when people are dependent on experts, the

100

degree of trust in the experts and consistency among them will have a strong impact on people's behavior. In crises involving complex technology, experts will not always agree and it is likely that, as during the TMI crisis, the media will emphasize those disagreements. Planners should therefore be prepared for the consequences of such disagreements, should they occur. Having a spokesperson who is seen as impartial and who is the primary source of technical information can help to maintain consistency in the information given to the public. This can help to avoid unnecessary public disagreements among experts, as happened early in the TMI crisis, when different spokespersons appeared to contradict each other because they had access to different information.

The strong fears about radiation, which were an important element in the TMI crisis, also have important implications for public planners. Evacuation during the crisis, as well as long-term psychological effects—that is, stress, coping, and attitudes—was influenced by people's fears of radiation. In future crises involving danger from radiation, officials should be aware of these strong feelings and how they are likely to affect people's behavior. It is also likely that dangers with characteristics similar to those of radiation (can spread over a large geographic area, cannot be detected without instruments, and with health effects that are not immediate) will have similar effects on evacuation and on long-term distress, as was found following the TMI accident.

The Three Mile Island experience also indicates that evacuation advisories issued by public officials can cause much larger evacuations than recommended, especially in situations where experts disagree and radiation or similar dangers are involved. At the same time, the TMI findings also showed that people did pay attention to what was said in the advisory, since the characteristics of people who evacuated were closely associated with the characteristics of those whom the governor advised to evacuate.

Planners will also want to take note of the .8% of people within five miles of TMI who said they did not evacuate because they were too sick or disabled to travel. While this percentage is small, it represents approximately two hundred people within a five-mile radius or thirty-two hundred within a fifteen-mile radius who would have required special medical assistance if a complete evacuation had been ordered. This would have presented public officials with a logistical problem of considerable magnitude.

Finally, the discrepancies between people's perceptions of the crisis's impact and objective evidence of that impact indicate that planners should rely, as much as possible, on objective data and not necessarily accept the views of the general public at face value when assessing the impact of a crisis. People's reports of their subjective feelings are, of course, valid in their own right and deserve consideration. However, these should be distinguished from their opinions about such topics as economic impact, where we found the public's views to be at variance with objective facts, and also from people's reports about the reasons for their behaviors, which, in the case of moving out of the area, we also found to be at variance with objective measures of their behaviors.

Implications for Social Scientists

In each of the previous chapters we have reviewed the relevant social science literature and discussed how the TMI findings we presented relate to that literature. For the most part, the findings we reported were consistent with the conceptualizations of earlier researchers. Two findings stand out as having important implications for future efforts to conceptualize how people respond to stressful situations. The first is that fear of possible harm can, by itself, have strong and long-lasting effects on people's feelings and behaviors. Previous research on how people respond to crises has typically been conducted in situations where there was both fear of harm and actual harm. In the TMI crisis there was no physical damage and no discernible short-term effects of the radiation released because of the accident. Therefore, the psychological and behavioral effects of the crisis were entirely due to fear of possible harm. The findings show that such fears can have substantial effects in their own right. As technological crises occur with greater frequency, fear reactions are likely to play an important role in how people respond.

The second finding with implications for social science theory is that certain coping responses were associated with maintaining high levels of distress rather than with decreasing distress. The explanation we propose for this finding is that people's coping efforts were unsuccessful in affecting the source of danger and therefore only served to remind them of their frustration and helplessness. This suggests that conceptualizations of coping should include the effectiveness of coping efforts in dealing with the problem. Where the situation is not responsive, coping efforts may lead to greater rather than less distress.

Finally, a question of importance to both planners and social scientists is the degree to which the Three Mile Island crisis is unique. For the planner, this question is important because the answer will indicate the degree to which the findings reported here can be used in planning for other types of crises. For the social scientist who studies how people respond to disasters, the answer will indicate the degree to which the same concepts can be applied across different crisis situations or whether different explanatory schemes are required for different types of crises, as, for example, suggested by Baum et al. (1983a).

In chapters 2 through 5 we have compared the TMI findings with those for other disasters and crises and found that the impact of the Three Mile Island crisis on persons living in the vicinity of the damaged reactor is largely consistent with what is known about responses to other types of stressful situations and therefore readily generalizable. Where we found differences they appeared to be related to three features of the TMI crisis: the technology of nuclear power production (its complexity, expense, and dangers), radiation as a source of danger (the difficulty of assessing danger without expert help, the potentially large geographic area that could be affected, and the fact that any health effects will not be known for a considerable period of time), and the tendency to blame those seen as responsible for the crisis (because human actions were seen as causing the danger). To the extent to which other crises share these characteristics, we predict that their effects will be similar to those observed following the TMI crisis and will be different from the effects of natural disasters in the same ways as were the impacts of the TMI crisis.

References

Adams, P. R., and G. R. Adams. "Mount Saint Helens's Ashfall. Evidence for a Disaster Stress Reaction," *Am Psychol* 39 (1984): 252–60.

Ahearn, F. L., Jr. "Disaster Mental Health: A Pre-Earthquake and Post-Earthquake Comparison of Psychiatric Admission Rates." *Urban Soc Chang Rev* 14 (1981): 22–28.

Ahearn, F. L., and R. E. Cohen. *Disasters and Mental Health: An Annotated Bibliography.* Washington, D.C.: National Institute of Mental Health, 1983.

Allee, D. *Trauma of the 1977 Flood in the Tug Fork Valley.* Ithaca: Dept. of Agricultural Economics, Cornell University, April 1980.

Andreoli, V., and S. Worchel. "Effects of Media, Communicator, and Message Position on Attitude Change." *Pub Opin Quart* 42 (1978): 59–70.

Andrews, G., C. Tennant, D. M. Hewson, and G. E. Vaillant. "Life Event Stress, Social Support, Coping Style, and Risk of Psychological Impairment." *J Nerv Ment Dis* 166 (1978): 307–16.

Aneshensel, C. S., and J. D. Stone. "Stress and Depression: A Test of the Buffering Model of Social Support." *Arch Gen Psychiatry* 39 (1982): 1392–96.

Antonovsky, A. *Health, Stress, and Coping.* San Francisco: Jossey-Bass, 1979.

Aguirre, B. E. "The Long Term Effects of Major Natural Disasters on Marriage and Divorce: An Ecological Study." *Victimology* 5 (1980): 298–307.

Babad, E. Y. "Effect of Source of Information as a Function of Age, Professional Relevance and Experience." *Psychol Rep* 41 (1977): 231–36.

Baker, G. W., and D. W. Chapman, eds. *Man and Society in Disaster.* New York: Basic Books, 1962.

Baldwin, B. A. "A Paradigm for the Classification of Emotional Crises: Implications for Crisis Intervention." *Am J Orthopsychiatry* 48 (1978): 538–51.

Balint, M. *The Doctor, His Patient, and the Illness.* New York: International Universities Press, 1957.

Bandura, A. "Self-Efficacy Mechanism in Human Agency." *Am Psychol* 37 (1982): 122–47.

Barnes, K., J. Brosius, S. L. Cutter, and J. K. Mitchell. "Human Responses by Impacted Populations to the Three Mile Island Nuclear Reactor Accident: An Initial Assessment." Unpublished manuscript, Dept. of Environmental Resources, Rutgers University, New Brunswick, N.J., 1979.

Barton, A. H. *Communities in Disaster. A Sociological Analysis of Collective Stress Situations.* Garden City: Doubleday, 1970, c1969.

Baum, A., R. J. Gatchel, and M. A. Schaeffer. "Emotional, Behavioral, and Physiological Effects of Chronic Stress at Three Mile Island." *J Consult Clin Psychol* 51 (1983): 565–72.

Baum, A., R. Fleming, and L. M. Davidson. "Natural Disaster and Technological Catastrophe." *Environ Behav* 15 (1983a): 333–54.

Baum, A., R. Fleming, and J. E. Singer. "Coping with Victimization by Technological Disaster." *J of Soc Issues* 39 (1983b): 117–38.

Bem, D. J. "Self-Perception Theory." In *Advances in Experimental Social Psychology,* vol. 6, ed. L. Berkowitz. New York: Academic Press, 1972.

Bennet, G. "Bristol Floods 1968. Controlled Survey of Effects on Health of Local Community Disaster." *Br Med J* 3 (1970): 454–58.

Berren, M. R., A. Beigel, and S. Ghertner. "A Typology for the Classification of Disasters: Implications for Intervention." *Community Ment Health J* 16 (1980): 103–11.

Bowen, R. M., R. P. Castanias, and L. A. Daley. "Intra-Industry Effects of the Accident at Three Mile Island." *J Fin and Quant Anal* 18 (1983): 87–111.

Broadhead, W. E., B. H. Kaplan, S. A. James, E. H. Wagner, V. J. Schoenbach, R. Grimson, S. Heyden, G. Tibblin, and S. H. Gehlbach. "The Epidemiologic Evidence for a Relationship Between Social Support and Health." *Am J Epidemiol* 117 (1983): 521–37.

Bromet, E. J., and L. Dunn. "Mental Health of Mothers Nine Months After the Three Mile Island Accident." *Urban Soc Chang Rev* 14 (1981): 12–15.

Bromet, E. J., L. Hough, and M. Connell. "Mental Health of Children near the Three Mile Reactor." *J Prev Psychiatry* 2 (1984): 275–301.

Bromet, E. J., D. K. Parkinson, H. C. Schulberg, L. O. Dunn, and P. C. Gondek. "Mental Health of Residents near the Three Mile Island Reactor: A Comparative Study of Selected Groups." *J Prev Psychiatry* 1 (1982a): 225–76.

Bromet, E., H. C. Schulberg, and L. Dunn. "Reactions of Psychiatric Patients to the Three Mile Island Nuclear Accident." *Arch Gen Psychiatry* 39 (1982b): 725–30.

Bromet, E. J., and H. C. Schulberg. "The TMI Disaster: A Search for High Risk Groups." In *Disaster Stress Studies: New Methods and Findings,* ed. J. H. Shore. American Psychiatric Press, Inc. Clinical Insights Monograph Series, 1986.

Brunn, S., J. Johnson, Jr., and D. Zeigler. *Final Report on a Social Survey of Three Mile Island Residents.* East Lansing: Dept. of Geography, Michigan State University, 1979.

Burke, J. D., Jr., J. F. Borus, B. J. Burns, K. H. Millstein, and M. C. Beasley. "Changes in Children's Behavior after a Natural Disaster." *Am J Psychiatry* 139 (1982): 1010–14.

Burnham, D. "The Press and Nuclear Energy." *Ann NY Acad Sci* 365 (1981): 107–9.

Chamberlin, B. C. "Mayo Seminars in Psychiatry: The Psychological Aftermath of Disaster." *J Clin Psychiatry* 41 (1980): 238–44.

Chan, K. B. "Individual Differences in Reactions to Stress and Their Personality and Situational Determinants: Some Implications for Community Mental Health." *Soc Sci Med* 11 (1977): 89–103.

Chapman, D. W. "Dimensions of Models in Disaster Behavior." In *Man and Society in Disaster,* ed. G. W. Baker and D. W. Chapman. New York: Basic Books, 1962.

Chisholm, R. F., S. V. Kasl, B. P. Dohrenwend, B. S. Dohrenwend, G. J. Warheit, R. L. Goldsteen, K. Goldsteen, and J. L. Martin. "Behavioral and Mental Health

Effects of the Three Mile Island Accident on Nuclear Workers: A Preliminary Report." *Ann NY Acad Sci* 365 (1981): 134–35.

Chisholm, R. F., S. V. Kasl, and L. Mueller. "The Effects of Social Support on Nuclear Worker Responses to the Three Mile Island Accident." *J Occup Behav* 7 (1986): 179–93.

Cleary, P. D. "Conceptualizing and Measuring Social Support." In *Evaluating Family Programs*, ed. H. B. Weiss and F. Jacobs. Chicago: Aldine, 1987.

Cleary, P. D., and R. Angel. "The Analysis of Relationships Involving Dichotomous Dependent Variables." *J Health Soc Behav* 25 (1984): 334–48.

Cleary, P. D., and P. S. Houts. "The Psychological Impact of the Three Mile Island Incident." *J Human Stress* 10 (1984): 28–34.

Clyne, M. B. *Night Calls: A Study in General Practice*. Philadelphia: Lippincott, 1961.

Cobb, S. "Social Support as a Moderator of Life Stress." *Psychosomatic Med* 38 (1976): 300–314.

Cohen, F., and R. S. Lazarus. "Coping and Adaptation and Health and Illness." In *Handbook of Health, Health Care, and the Health Professions*, ed. D. Mechanic. New York: The Free Press, 1983.

Cohen, S., and G. McKay. "Interpersonal Relationships as Buffers of the Impact of Psychological Stress on Health." In *Handbook of Psychology and Health*, ed. A. Baum, J. E. Singer, and S. E. Taylor. Hillsdale, N.J.: Erlbaum, 1984.

Cohen, S., and S. L. Syme, eds. *Social Support and Health*. New York: Academic Press, 1985.

Collins, D. L., A. Baum, and J. E. Singer. "Coping with Chronic Stress at Three Mile Island: Psychological and Biochemical Evidence." *Health Psychol* 2 (1983): 149–66.

Commonwealth of Pennsylvania, Governor's Office of Policy and Planning. *The Socioeconomic Impacts of the Three Mile Island Accident*. Harrisburg, December 1979.

Davidson, L. M., and A. Baum. "Chronic Stress and Posttraumatic Stress Disorders." *J Consult Clin Psychol* 54 (1986): 303–8.

Davidson, L. M., A. Baum, and D. L. Collins. "Stress and Control-Related Problems at Three Mile Island." *J Appl Soc Psychol* 12 (1982): 349–59.

Davis, F. *Passage Through Crisis*. Indianapolis: Bobbs-Merrill, 1963.

Dew, M. A., E. J. Bromet, and H. Schulberg. "Application of a Temporal Persistence Model to Community Residents' Long-Term Beliefs about the Three Mile Island Nuclear Accident." *J Appl Soc Psychol* (in press).

Dew, M. A., E. J. Bromet, and H. C. Schulberg. "A Comparative Analysis of Two Community Stressors' Long-Term Mental Health Effects." *Am J Community Psychol* 15 (1987a): 167–84.

Dew M. A., E. J. Bromet, H. C. Schulberg, L. O. Dunn, and D. K. Parkinson. "Mental Health Effects of the Three Mile Island Nuclear Reactor Restart." *Am J Psychiatry* 144 (1987b): 1074–77.

Dohrenwend, B. P., B. S. Dohrenwend, S. V. Kasl, and G. J. Warheit. *Technical Staff Analysis Report on Behavioral Effects to the President's Commission on the Accident at Three Mile Island*. Washington, D.C.: Government Printing Office, 1979.

Dohrenwend, B. P., B. S. Dohrenwend, G. J. Warheit, G. S. Bartlett, R. L. Goldsteen,

K. Goldsteen, and J. L. Martin. "Stress in the Community: A Report to the President's Commission on the Accident at Three Mile Island." *Ann NY Acad Sci* 365 (1981): 159–74.

Dohrenwend, B. P., P. E. Shrout, G. Egri, and F. S. Mendelsohn. "Nonspecific Psychological Distress and Other Dimensions of Psychopathology. Measures for Use in the General Population." *Arch Gen Psychiatry* 37 (1980): 1229–36.

Drabek, T. E. "Methodology of Studying Disasters." *Am Behav Scientist* 13 (1970): 331–43.

Drayer, C. S. "Psychological Factors and Problems, Emergency and Long-Term." *Ann Am Acad Pol Soc Sci* (1957): 151–59.

Dynes, R. R., and E. L. Quarantelli. *Organizational Communications and Decision Making in Crises.* Disaster Res. Cent. Misc. Rep. No. 18. Columbus: Ohio State Univ., 1976.

Erikson, K. T. "Disaster at Buffalo Creek. Loss of Communality at Buffalo Creek." *Am J Psychiatry* 133 (March 1976a): 302–5.

Erikson, K. T. *Everything in Its Path. Destruction of Community in the Buffalo Creek Flood.* New York: Simon and Schuster, 1976b.

Evans, N. "Hidden Costs of the Accident at Three Mile Island." *Energy* 7 (1982): 723–30.

Fabrikant, J. I. "The Effects of the Accident at Three Mile Island on the Mental Health and Behavioral Responses of the General Population and Nuclear Workers." *Health Phys* 45 (1983): 579–86.

Festinger, L. *A Theory of Cognitive Dissonance.* Stanford: Stanford University Press, 1957.

Fienberg, S. E., E. J. Bromet, D. Follmann, D. Lambert, and S. M. May. "Longitudinal Analysis of Categorical Epidemiological Data: A Study of Three Mile Island." *Environ Health Perspect* 63 (1985): 241–48.

Firebaugh, M. W. "Public Attitudes and Information on the Nuclear Option." *Nuclear Safety* 22 (1981): 147–56.

Fleming, R., A. Baum, M. M. Gisriel, and R. J. Gatchel. "Mediating Influences of Social Support on Stress at Three Mile Island." *J Human Stress* 8 (1982): 14–22.

Flynn, C. B. *Three Mile Island Telephone Survey, Preliminary Report on Procedures and Findings.* Seattle: Social Impact Research, Inc., September 1979.

Frank, J. D. *Persuasion and Healing.* Baltimore: The Johns Hopkins Press, 1973.

Frederick, C. J. "Current Thinking About Crisis or Psychological Intervention in United States Disasters." *Mass Emerg* 2 (1977): 43–50.

Fruedenburg, W. R., and R. K. Baxter. "Host Community Attitudes toward Nuclear Power Plants: A Reassessment." *Soc Sci Quart* 65 (1984): 1129–36.

Gamble, H. B., and R. H. Downing. *Effects of the Accident at Three Mile Island on Residential Property Value and Sales.* University Park: Institute for Research on Land and Water Resources, Pennsylvania State University, April 1981.

Gatchel, R. J., M. A. Schaeffer, and A. Baum. "A Psychophysiological Field Study of Stress at Three Mile Island." *Psychophysiology* 22 (1985): 175–81.

Gibbs, L. M. "Community Response to an Emergency Situation: Psychological Destruction and the Love Canal." *Am J Community Psychol* 11 (1983): 116–25.

Glenn, C. M. "Natural Disasters and Human Behavior: Explanation, Research, and Models." *Psychology: Quart J Human Behav* 16 (1979): 23–36.

Goldhaber, M. K., P. S. Houts, and R. DiSabella. "Moving after the Crisis: A Prospective Study of Three Mile Island Population Mobility." *Environ Behav* 15 (1983): 93–120.

Goldsteen, R., and J. K. Schorr. "The Long-Term Impact of a Man-Made Disaster: An Examination of a Small Town in the Aftermath of the Three Mile Island Nuclear Reactor Accident." *Disasters* 6 (1982): 50–59.

Green, B. L. "Assessing Levels of Psychological Impairment Following Disaster: Consideration of Actual and Methodological Dimensions." *J Nerv Ment Dis* 170 (1982): 544–52.

Hall, P. S., and P. W. Landreth. "Assessing Some Long-Term Consequences of a Natural Disaster." *Mass Emerg* 1 (1975): 55–61.

Hans, J. M., Jr., and T. C. Sell. *Evacuation Risks—An Evaluation.* Las Vegas: U.S. Environmental Protection Agency, 1974.

Hartsough, D. M., and J. C. Savitsky. "Three Mile Island. Psychology and Environmental Policy at a Crossroads." *Am Psychol* 10 (1984): 1113–22.

Holden, C. "Love Canal Residents Under Stress." *Science* 208 (1980): 1242–44.

House, J. S. *Work Stress and Social Support.* Reading, Mass.: Addison-Wesley, 1981.

House, J. S., and R. L. Kahn. "Measures and Concepts of Social Support." In *Social Support and Health,* ed. S. Cohen and L. Syme. New York: Academic Press, 1984.

Houts, P. S., T. W. Hu, R. A. Henderson, P. D. Cleary, and G. Tokuhata. "Utilization of Medical Care Following the Three Mile Island Crisis." *Am J Public Health* 74 (1984): 140–42.

Houts, P. S., M. K. Lindell, T. W. Hu, P. D. Cleary, G. Tokuhata, and C. B. Flynn. "The Protective Action Decision Model Applied to Evacuation During the Three Mile Island Crisis." *Int J Mass Emerg Disasters* 2 (1984): 27–39.

Houts, P. S., R. W. Miller, and M. Lindell. "Reanalysis of Three Mile Island Telephone Survey Data." Submitted in partial fulfillment of contract RFQ No. RS-NRR-82-123. Nuclear Regulatory Commission, Washington, D.C., 1983.

Houts, P. S., R. M. DiSabella, and M. K. Goldhaber. "Health-Related Behavioral Impact of the Three Mile Island Nuclear Incident, Part III." Report submitted to the TMI Advisory Panel on Health Research Studies of the Pennsylvania Department of Health, Harrisburg, 1981.

Houts, P. S., and M. K. Goldhaber. "Psychological and Social Effects on the Population Surrounding Three Mile Island After the Nuclear Accident on March 28, 1979." In *Energy, Environment and the Economy,* ed. S. Majuindar. Pennsylvania Academy of Sciences, 1981.

Houts, P., R. W. Miller, G. K. Tokuhata, and K. S. Ham. "Health-Related Behavioral Impact of the Three Mile Island Nuclear Incident." Report submitted to the TMI Advisory Panel on Health Research Studies of the Pennsylvania Department of Health, Harrisburg, 1980.

Hu, T. W., and K. S. Slaysman. "Health-Related Economic Costs of the Three Mile Island Accident." *Socio-Econ Plan Sci* 18 (1984): 183–93.

Huerta, F., and R. Horton. "Coping Behavior of Elderly Flood Victims." *Gerontologist* 18 (1978): 541–46.

Jaeger, M. E., S. Anthony, and R. L. Rosnow. "Who Hears What from Whom and with What Effect: A Study of Rumor." *Person Soc Psychol Bull* 6 (1980): 473–78.

Johnson, J. H., Jr., and D. J. Zeigler. "Modelling Evacuation Behavior during the Three Mile Island Crisis." *Socio-Econ Plan Sci* 20 (1986): 165–71.

Kaplan, H. B. "Self-derogation and Adjustment to Recent Life Experiences." *Arch Gen Psychiatry* 1 (1970): 324–31.

Kasl, S. V., R. F. Chisholm, and B. Eskenazi. "The Impact of the Accident at Three Mile Island on the Behavior and Well-Being of Nuclear Workers. Part II: Job Tension, Psychophysiological Symptoms, and Indices of Distress." *Am J Public Health* 71 (1981): 484–95.

Katz, J. L., H. Weiner, T. F. Gallagher, and L. Hellman. "Stress, Distress, and Ego Defenses. Psychoendocrine Response to Impending Breast Tumor Biopsy." *Arch Gen Psychiatry* 23 (1970): 131–42.

Killian, L. M. "Some Accomplishments and Some Needs in Disaster Study." *J of Soc Issues* 10 (1954): 66–72.

Kinston, W., and R. Rosser. "Disaster: Effects on Mental and Physical State." *J Psychosom Res* 18 (1974): 437–56.

Kobasa, S. C., S. R. Maddi, and S. Courington. "Personality and Constitution as Mediators in the Stress-illness Relationship." *J Health Soc Behav* 22 (1981): 368–78.

Kohn, M. L. "The Interaction of Social Class and Other Factors in the Etiology of Schizophrenia." *Am J Psychiatry* 133 (1976): 177–80.

Kohn, M. L. *Class and Conformity: A Study of Values.* 2d ed. Chicago: Univ. of Chicago Press, 1977.

Kraybill, D. *Three Mile Island: Local Residents Speak Out.* Elizabethtown, Pa.: The Social Research Center, 1979.

Kraybill, D. *Three Mile Island: Local Residents Speak Out Twice.* Elizabethtown, Pa.: The Social Research Center, 1980.

Lazarus, R. S. "The Stress and Coping Paradigm." In *Models for Clinical Psychopathology,* ed. C. Eisendorfer, D. Cohen, A. Kleinman, and P. Maxim. New York: Spectrum, 1981.

Lefcourt, H. M. *Locus of Control: Current Trends in Theory and Research.* New York: Halstead, 1976.

Lefcourt, H. M. "Locus of Control and Stressful Life Events." In *Stressful Life Events and Their Contexts,* ed. B. S. Dohrenwend and B. P. Dohrenwend. New York: Prodist, 1981.

Leventhal, H., M. A. Safer, and D. M. Panagis. "The Impact of Communications on the Self-Regulation of Health Beliefs, Decisions, and Behavior." *Health Educ Q* 10 (1983): 3–29.

Levine, A. G. *Love Canal: Science, Politics, and People.* Lexington, Mass.: Lexington Books, 1982.

Lifton, R. J., and E. Olson. "The Human Meaning of Total Disaster. The Buffalo Creek Experience." *Psychiatry* 39 (1976): 1–18.

Lindemann, E. "Symptomatology and Management of Acute Grief." *Am J Psychiatry* 101 (1944): 141–48.

Lipowski, Z. J. "Physical Illness, the Individual and the Coping Processes." *Psychiatry Med* 1 (1970): 91–102.

Logue, J. N., and H. Hansen. "A Case-Control Study of Hypertensive Women in a Post-Disaster Community: Wyoming Valley, Pennsylvania." *J Human Stress* 6 (1980): 28–34.

Logue, J. N., M. E. Melick, and E. L. Struening. "A Study of Health and Mental Health Status Following a Major Natural Disaster." *Res Commun Ment Health* 2 (1981): 217–74.

Malinowski, B. *Magic, Science and Religion and Other Essays.* Garden City, N.Y.: Anchor Books, 1955.

Marshall, E. "Public Attitudes to Technological Progress." *Science* 205 (1979): 281, 283–85.

Mazur, A. "Commentary: Opinion Poll Measurement of American Confidence in Science." *Sci Tech Human Values* 6 (1981): 16–19.

Mechanic, D. "Social Psychological Factors Affecting the Presentation of Bodily Complaints." *N Engl J Med* 286 (1972): 1132–39.

Melick, M. E. "Social, Psychological and Medical Aspects of Stress-related Illness in the Recovery Period of a Natural Disaster." Ph.D. dissertation, State University of New York at Albany (#76-18): p. 431 (1976).

Melick, M. E. "Life Change and Illness: Illness Behavior of Males in the Recovery Period of a Natural Disaster." *J Health Soc Behav* 19 (1978a): 335–42.

Melick, M. E. "Self-Reported Effects of a Natural Disaster on the Health and Well-Being of Working Class Males." *Crisis Intervention* 1 (1978b): 12–31.

Melick, M. E., J. N. Logue, and C. J. Frederick. "Stress and Disaster." In *Handbook of Stress: Theoretical and Clinical Aspects,* ed. L. Goldberger and S. Breznitz. New York: Free Press, 1982.

Menninger, W. C. "Psychological Reactions in an Emergency." *Am J Psychiatry* 109 (1952).

Mileti, D. S., D. M. Hartsough, P. Madson, and R. Hufnagel. "The Three Mile Island Incident: A Study of Behavioral Indicators of Human Stress." *Int J Mass Emerg Disasters* 2 (1984): 89–113.

Mileti, D. *Natural Hazard Warning Systems in the United States: A Research Assessment.* Boulder: Institute of Behavioral Science, University of Colorado, 1975.

Mills, S. "Voices from Three Mile Island." Edited collection in *The Progressive* 44 (1980): 16–24.

Milne, G. "Cyclone Tracey: I. Some Consequences of the Evacuation for Adult Victims." *Aust Psychol* 12 (1977a): 39–54.

Milne, G. "Cyclone Tracey: II. The Effects on Darwin Children." *Aust Psychol* 12 (1977b): 55–62.

Mitchell, R. C. "Polling on Nuclear Power: A Critique of the Polls After Three Mile Island." In *Polling on the Issues,* ed. A. H. Cantril. Washington, D.C.: Seven Locks Press, 1980.

Mitchell, R. C. "Public Response to a Major Failure of a Controversial Technology." In *Accident at Three Mile Island: The Human Dimensions,* ed. D. L. Sills, C. P. Wolf, and V. B. Shelanski. Boulder: Westview Press, 1982.

Moore, H. E. "Some Emotional Concomitants of Disaster." *Ment Hyg* 42 (1958): 45–50.

111

Moore, H. E., and H. J. Friedsam. "Reported Emotional Stress Following a Disaster." *Social Forces* 38 (1959): 135–39.

Moos, R. H., and V. Tsu. "The Crisis of Physical Illness: An Overview." In *Coping with Physical Illness*, ed. R. H. Moos. New York: Plenum Press, 1977.

National Center for Health Statistics. *Acute Conditions Incidence and Associated Disability*. DHEW Publication No. (PHS) 79-2560, 1979.

Nealy, S. M., B. D. Melber, and W. L. Rankin. *Public Opinion and Nuclear Energy*. Lexington, Mass.: Lexington Books, 1983.

Nelson, J. P. "Three Mile Island and Residential Property Values: Empirical Analysis and Policy Implications." *Land Econ* 57 (1981): 363–72.

Nesterenko, A., and D. L. Eckberg. "Attitudes Toward Technology and Science in a Biblebelt City." *Free Inquiry Creat Soc* 12 (1984).

Newman, C. J. "Children of Disaster: Clinical Observations at Buffalo Creek." *Annu Prog Child Psychiatr Child Dev* 10 (1977): 149–61.

Nimmo, D. "TV Network News Coverage of Three Mile Island: Reporting Disasters as Technological Fables." *Int J Mass Emerg Disasters* 2 (1984): 115–45.

Ollendick, D. G., and M. Hoffman. "Assessment of Psychological Reactions in Disaster Victims." *J Community Psychol* 10 (1982): 157–67.

Parkinson, D. K., and E. J. Bromet. "Correlates of Mental Health in Nuclear and Coal-Fired Power Plant Workers." *Scand J Work Environ Health* 9 (1983): 341–45.

Pearlin, L. I., and C. Schooler. "The Structure of Coping." *J Health Soc Behav* 19 (1978): 2–21.

Penick, E. C., B. J. Powell, and W. A. Sieck. "Mental Health Problems and Natural Disaster: Tornado Victims." *J Community Psychol* 4 (1976): 64–67.

Perry, R. W. *Citizen Evacuation in Response to Nuclear and Nonnuclear Threats*. Seattle: Battelle Human Affairs Research Centers, September 1981.

Perry, R. W., and M. K. Lindell. "The Psychological Consequences of a Natural Disaster. A Review of Research on American Communities." *Mass Emerg* 3 (1978).

Perry, R. W., M. K. Lindell, and M. R. Greene. *Evacuation Planning in Emergency Management*. Lexington, Mass.: Lexington Books, 1981.

President's Commission on the Accident at Three Mile Island. *The Need for Change: The Legacy of TMI*. Washington, D.C.: Government Printing Office, 1979.

Prince-Embury, S., and J. Rooney. "Psychological Symptoms of Middletown Residents Post-Restart of Three Mile Island." Paper presented at American Psychological Association meetings, August 1987.

Quarantelli, E. L., ed. *Disasters: Theory and Research*. London: Sage, 1977

Quarantelli, E. *Evacuation Behavior and Problems: Findings and Implications from the Research Literature*. Columbus: Disaster Research Center, Ohio State University, 1980.

Quarantelli, E. L., and R. R. Dynes. "Response to Social Crisis and Disaster." *Annu Rev Soc* 3 (1977): 23–49.

Rangell, L. "Disaster at Buffalo Creek. Discussion of the Buffalo Creek Disaster: The Course of Psychic Trauma." *Am J Psychiatry* 133 (1976): 313–16.

Reko, K. "The Psychological Impact of Environmental Disasters." *Bull Environ Contam Toxicol* 33 (1984): 655–61.

Richard, W. C. "Crisis Intervention Services Following Natural Disaster: The Pennsylvania Recovery Project." *J Community Psychol* 2 (1974): 211–19.

Rosenberg, M. *Conceiving the Self.* New York: Basic Books, 1979.

Rosnow, R. L. "Psychology of Rumor Reconsidered." *Psychol Bull* 87 (1980): 578–91.

Roth, S., and L. J. Cohen. "Approach, Avoidance and Coping with Stress." *Am Psychol* 41 (1986): 813–19.

Rubin, D. M. "What the President's Commission Learned About the Media." *Ann NY Acad Sci* 365 (1981): 95–106.

Ruskin, A., O. W. Beard, and R. L. Schaffer. "Blast Hypertension." *Am J of Med* 4 (1948): 228–36.

Sandman, P. M., and M. Paden. "At Three Mile Island." *Columbia Journ Rev* 18 (1979): 43–58.

Schaeffer, M. A., and A. Baum. "Adrenal Cortical Response to Stress at Three Mile Island." *Psychosom Med* 46 (1984): 227–37.

Schorr, J. K., R. Goldsteen, and C. H. Cortes. "The Long-Term Impact of a Man-Made Disaster: A Sociological Examination of a Small Town in the Aftermath of the Three Mile Island Nuclear Reactor Accident." Presented at the Tenth World Congress of Sociology, Mexico City, August 1982.

Seligman, M. E. P. *Helplessness.* San Francisco: W. H. Freeman, 1975.

Selye, H. *The Stress of Life.* New York: McGraw-Hill, 1956.

Silver, R. L., and C. B. Wortman. "Coping with Undesirable Life Events." In *Human Helplessness: Theory and Applications,* ed. J. Garber and M. E. P. Seligman. New York: Academic Press, 1980.

Smith, M. H. "The Three Mile Island Evacuation: Voluntary Withdrawal from a Nuclear Power Plant Threat." Unpublished paper, Long Island University, C. W. Post Center, Department of Sociology, 1979.

Smith, R. E., J. H. Johnson, and I. G. Sarason. "Life Change, the Sensation Seeking Motive, and Psychological Distress." *J Consult Clin Psychol* 46 (1978): 348–49.

Solomon, Z. "Stress, Social Support and Affective Disorders in Mothers of Pre-School Children—A Test of the Stress-Buffering Effect of Social Support." *Soc Psychiatry* 20 (1985): 100–105.

Solomon, Z., and E. Bromet. "The Role of Social Factors in Affective Disorder: An Assessment of the Vulnerability Model of Brown and His Colleagues." *Psychol Med* 12 (1982): 123–30.

Sorensen, J., J Soderstrom, E. Copenhaver, S. Carnes, and R. Bolin. *Impacts of Hazardous Technology, the Psychosocial Effects of Restarting TMI-1.* Albany: State University of New York Press, 1987.

Stallings, R. A. "Evacuation Behavior at Three Mile Island." *Int J Mass Emerg Disasters* 2 (1984): 11–26.

Starr, P., and W. Pearman. *Three Mile Island Sourcebook, Annotations of a Disaster.* New York: Garland, 1983.

Streufert, S., R. Pogash, M. Piasecki, A. Baum, J. Logue, K. Sivarajah, and R. Strohman. "Health Consequences of a Chronic Human-made Disaster." Unpublished manuscript, 1987.

Sudman, S. "Telephone Methods in Survey Research: The State of the Art." Presented to American Educational Research Association, Los Angeles, April 14, 1981.

Taylor, V. "Good News about Disaster." *Psychol Today* 11 (1977): 93+.

Taylor, V. A., G. A. Ross, and E. L. Quarantelli. "Delivery of Mental Health Services in Disasters." Columbus: Disaster Research Center, Ohio State University, 1976.

Tierney, K. J., and B. Baisden. "Crisis Intervention Programs for Disaster Victims: A Source Book and Manual for Smaller Communities." DHEW Publication No. (ADM) 79-675, 1979.

Titchener, J. L., and F. T. Kapp. "Disaster at Buffalo Creek. Family and Character Change at Buffalo Creek." *Am J Psychiatry* 133 (1976): 295–99.

Upton, A. C. "Health Impact of the Three Mile Island Accident." *Ann NY Acad Sci* 365 (1981): 63–75.

U.S. Senate Hearings. *Disaster Assistance Pacific Northeast—Mount Saint Helens Eruption.* Washington, D.C.: Committee of Appropriations, 1980.

Value Line Investment Survey. New York: Value Line Inc., 1987.

Van der Pligt, J., J. R. Eiser, and R. Spears. "Construction of a Nuclear Power Station in One's Locality: Attitudes and Salience." *Basic Appl Soc Psychol* 7 (1986a): 1–15.

Van der Pligt, J., J. R. Eiser, and R. Spears. "Attitudes toward Nuclear Energy: Familiarity and Salience." *Environ Behav* 18 (1986b): 75–93.

Walsh, E. J. "Resource Mobilization and Citizen Protest in Communities Around Three Mile Island." *Social Problems* 29 (1981): 1–21.

Walsh, E. J. "The Role of Target Vulnerabilities in High Technology Protest Movements: The Nuclear Establishment at Three Mile Island." *Sociol Forum* 1 (1986): 199–218.

Walsh, E. J., and R. H. Warland. "Social Movement Involvement in the Wake of a Nuclear Accident: Activists and Free Riders in the TMI Area." *Am Sociol Rev* 48 (1983): 764–80.

Wheaton, B. "Stress, Personal Coping Resources, and Psychiatric Symptoms: An Investigation of Interactive Models." *J Health Soc Behav* 24 (1983): 208–29.

White, G. F., and J. E. Haas. *Assessment of Research on Natural Hazards.* Cambridge: MIT Press, 1975.

Wright, J. D., and P. H. Rossi, eds. *Social Science and Natural Hazards.* Cambridge: Abt Books, 1981.

Wright, J. D., P. H. Rossi, S. R. Wright, and B. E. Weber. *After the Cleanup. Long Range Effects of Natural Disasters.* Beverly Hills: Sage, 1979.

Zeigler, D. J., S. D. Brunn, and J. H. Johnson, Jr. "Evacuation from a Nuclear Technological Disaster." *Geogr Rev* 71 (1981): 1–16.

Index

116